# Thriving Cities: A Guide to Sustainable Urban Development for Flourishing Communities

Chris

**Table of Contents**

# CHAPTER 1: INTRODUCTION

"What is the city but the people?" — Shakespeare

By 2030, it is estimated that over 70% of the global community will live in cities. Stakeholders in their cities and communities are increasingly concerned with how their development initiatives reconfigure their city's social, economic, political, and ecological system resources towards a more sustainable future. As a post post-modernist concern, our relationships with the natural environment and social well-being of societies and socio-economic environments, are becoming increasingly apparent as metrics for community-wide sustainability and whole systems flourishing. The meaning of sustainability in the context of this book is more aligned with flourishing, and flourishing can be expressed as panarchic thriving. Flourishing focuses on increasing connectedness, prosperity for all, contributing to a regenerative natural environment, and improving human well-being (Laszlo, 2020). This definition of "sustainable-as-flourishing" is different from the well-accepted traditional translation of sustainability understood as a reductive process framework of reducing harm, ecological footprint, and social inequities, while my definition of flourishing stems from a positive psychological and its generative process. Throughout this book, I combine the terms "sustainability" and "flourishing," but to minimize confusion of the contextual meaning, I use the term "sustainability" to be interpreted as "sustainability-as-flourishing."

Measuring city-level performance and implementing concrete sustainability development initiatives toward flourishing cities are among the biggest challenges societies face in the first quarter of the twenty-first century. According to the United

Nation's Cities Sustainability Development Report, there has been some significant progress towards achieving sustainability goals in some cities, while other cities have remained static; or worse, have declined in a downward spiral (Lynch, 2019; Prakash, Teksoz, Espey, Sachs, Shank, & Schmidt-Traub, 2017). Human interactions such as capacity building, innovation, adaptation-reflexivity dynamics, and materializations are critical characteristics that are at the core of determining a city's thriving vitality. Additionally, how our societies successfully manage natural resources, political influence, social well-being, economic prosperity, and the built and cultural environments as a whole are even more critical elements that can contribute to a flourishing vitality for cities.

Over the last three decades, scholars have observed how institutional, organizational, and political leaders and peripheral stakeholders have approached sustainability in their respective cities by reductionists and fragmented methodologies (McQuaid, 2019). As a result, these efforts have translated into modest to poor community performance indicators, slow incremental change, and bolted-on sustainability initiatives that have not been sufficient to reverse the effects of climate change and simultaneously promote in mass social prosperity as a sustainable city. In addition to sustainability goal attainment, a transformational paradigm shift towards a consciousness of interconnectedness is needed for all stakeholders in cities to achieve a holistic flourishing goal of a City as a Sustainable Ecosystem (Cooperrider & McQuaid, 2012; Newman & Jennings, 2012; Romanelli, 2018). Cities as sustainable ecosystems can be described as urban development and design that integrates and aligns human and non-human interactions and resource flows with the Earth's natural cycles. Sustainability principles and practices are embedded throughout the ecosystem. However, in order to achieve the challenges of transforming city ecosystems to sustainable and flourishing, require contemporary leaders and managers a holistic lens to approach sustainability with an integrated whole-system, humanistic strategy. In his seminal work, "Urban Dynamics," Forrester (1970) suggested that a holistic approach may be more effective for handling the complexities of urban development. While there has been some resistance against sustainable development and social flourishing, there are good reasons for the multitude of stakeholders in city ecosystems and the business and political institutions to adopt a holistic integrative approach instead of aggregated reductionist approaches to sustainable development. First, understanding and embracing the complexities of ecosystems provide leaders and managers a perspective to cope with the growing need for quality management integration approaches (Kay & Schneider, 1995) with a level of uncertainty (Prigogine & Stengers, 1997; Stacey, 2002) and to deal with emergent new dynamic socio-ecological patterns within these ecosystems (Fernández, Maldonado, & Gershenson, 2014; Holling, 2001). Integrated holistic approaches are also aligned with contemporary concerns for plurality and diversity (Schoenberger-Orgad & McKie, 2005) in organizations and communities. Second, the diversity of stakeholders and macro-managers in city development have indicated sustainable social, architectural, infrastructure, and artifacts innovations are needed to create flourishing cities and communities on a global scale (Berkowitz, 2018; Cherkowski & Walker, 2014; Hart, 2017). Third, the radical shift needed for both nature and society to flourish will most likely not emerge from a single actor, organization, institution, nor city-nation. This will require a conscious collective action to innovate and implement change and maintain a resilient city as an ecosystem at the scale of the whole (Bec, Char-lee, & Moyle, 2019;

Cooperrider & McQuaid, 2012; Dale, Ling, & Newman, 2010; Ostrom, 2010). Fourth,

significant empirical work has observed integrative transitions towards sustainable

ecosystems, communities, and industry sectors. Studies have presented this transition as a

manifestation of ecological complex adaptive systems (Holling, 2001) using scenario-

based simulation modeling (Peter & Swilling, 2014) that is implemented using multi-

stakeholder processes (Bäckstrand, 2006) that deploy macro-management approaches

(Cooperrider & McQuaid, 2012; McFarland, 1977; Popkov, 1984; Sperry, 1993) that

form meta-organizational structures and capabilities (Berkowitz, 2018; Valente & Oliver,

2018) and develop related dynamic capabilities (Heaton, Siegel, & Teece, 2019).

There is an opportunity to fill a gap in the literature that explicates a framework of

macro-management and dynamic capabilities through the lens of city and community

sustainability and flourishing transformation. My empirical motivation is to extend the

literature on integrative ecosystem management approaches which can transform cities as

sustainable ecosystems with a flourishing societal vitality. This leads to the primary

research question of my book:

> *What are integrative ecosystems management approaches, and how are they*
> *instrumental in the development of flourishing cities as sustainable ecosystems*
> *(CASE)?*

To support my inquiry with empirical evidence, I utilized an integrative meta-theory and

employed a mixed methods approach. The mixed methods approach was appropriate for

this research because I wanted to study different dimensions of sustainable development

in city ecosystems and provide empirical evidence to support my claim that integrative

ecosystem management approaches can be instrumental in the development of flourishing cities as sustainable ecosystems. The benefit of a mixed methods approach is that it provides multiple aspects of the same phenomena to advance my theory of ecosystem sustainable development. Studying a meta-theoretical framing of ecosystem sustainable development integrated with the nested meta-organization, macromanagement, and dynamic capabilities frameworks seeks to understand the emerging concerns and problem domains by building, applying, and studying the effects of multi-stakeholder processes and the implementing organization for sustainable development in city ecosystems. Not only do these nested frameworks provide empirical evidence from prior empirical work, but my three subsequent research studies were positioned to provide additional empirical evidence in supporting my unit theories, which in turn, also supports the integrated meta-theory of sustainable development. The mixed methods approach was beneficial in studying this phenomenon because it helped solve the internal-external paradox of a city ecosystem (whole) and its embedded panarchic systems (parts). Moreover, my three subsequent research questions were positioned to support different aspects of my primary research question (Teddlie & Tashakkori, 2006). I conducted the three studies to explore integrative ecosystems management (IEM) factors, which led to flourishing cities as sustainable ecosystems, and how implementing organizations utilize IEM for ecosystem development performance. My studies sought to inform: What forms integrative ecosystem management approaches, what has been successful in sustainable development, how do Appreciative Inquiry (AI) platforms support sustainable innovation and development, and the combinations of factors that led to high sustainable development performance. The first study observed antecedents that led to successful sustainable development initiatives in cities. The second study applies a structural equation modeling SEM for measuring and validating the effects of a city and community-level macromanagement approach towards building innovation capacity and resilience. The third study uses a fuzzy-set qualitative comparative analysis (fsQCA) to model the effects of combinations of multi-level factors on the performance of a city's ecosystem sustainable development goal (SDG) achievement (Fiss, 2011).

The first empirical work was a qualitative study using grounded theory. The inquiry sought the emergence of factors that led to successful sustainable development implementation in cities. Targeted from the 2017 Sustainable Development Solution Network's United Nations Sustainable Development Goals City Report (Prakash et al., 2017), I chose four low performing cities (according to total city index scores), then asked leaders, managers, stakeholders in business communities, government sectors, non-profit organizations, and activists their lived experiences of implementing sustainable development initiatives. This study aimed to discover the factors that lead to successful sustainable development in cities. The thematic categories that emerged from the coding process were capacity building, adaptive learning, collaboration, and a multi-stakeholder governance process. My interview results also revealed a lack of successful sustainable development initiatives designed for the social well-being indicators of sustainability. Additionally, this indicated a deficiency of deep whole-systems collaboration and validated my claim that fragmented, reductionist approaches to city development remain dominant strategies that leaders, managers, and non-profit organizations in most cities follow.

The second empirical work was a quantitative study using SEM. The constructs developed in this causal model included Appreciative Inquiry (AI), the strengths,

opportunities, aspiration, results (SOAR) framework, innovative capacity, and resilience.
The AI platform coupled with that SOAR framework was developed as an example of the
macro-management approach. I hypothesized a model to measure and test the AI-SOAR
operationalized system and process effectiveness on cities as a sustainable ecosystem for
innovation and resilience. I surveyed business leaders, government officials, managers,
participative citizens, and general stakeholders in various U.S. cities at random.
Qualitative research from academics and practitioners has proven AI platforms'
effectiveness for systematic change, and the causal model was designed as a quantitative
measurement tool to test the impact of this macromanagement approach for ecosystem
transformation towards flourishing cities. My results revealed that AI-SOAR has a
positive effect on innovative capacity building and resiliency, promoting a flourishing
ecosystem vitality. It also validated the double mediating effects AI-SOAR leads to
innovative capacity on resiliency and resiliency on innovative capacity. This is consistent
with the empirical work on the paradoxical dynamics between innovation and resiliency
detailed by Dale et al. (2010).

The third empirical work is a fuzzy-set qualitative comparative case analysis
(Ragin, 2014) of city ecosystems' sustainable development performance progress towards
sustainable cities. I offer several distinctions between the ecosystem cases and analyze
how stakeholders in cities approach sustainable development differently and what are
related outcomes. I propose this comparative analysis will identify combinations of
factors that can lead to sustainable development achievement total score improvements
(or their decline) (based on the Annual UN SDG 17 City Report). I utilized the data
collected from my prior studies' interviews and secondary data collected over a three-

year period. The results will validate those cities' ecosystems applying a macromanagement approach and going beyond community operational development by employing dynamic capabilities within their approach will have yielded better outcomes towards the transformation of sustainable cities with mass flourishing.

As a continuum of this book, I will first highlight the relevant literature stream from which my theories of integrative ecosystem management, macromanagement, meta-organization, and dynamic capabilities are informed. This will set the theoretical framework. Then I will discuss the research agenda and design, as well as the reasoning why I chose those methodologies. Next, the three studies with the inclusion of theoretic frameworks, methods, and results will be covered. Following, I offer the deliberation of my integrated results and discussion with implications to theory and practice from my findings. Finally, I close with a conclusion and an agenda for future research opportunities.

**CHAPTER 2: THEORETICAL FRAMING**

This book draws upon the theories and practice of formal ontologies in complex systems approaches to sustainability and meta-organizational behaviors within positive organizational scholarship (POS) related to the collective human, organizational, and institutional interactions nested within complex adaptive and responsive systems of a city. The human organizing structures, governance, and processes within a larger complex system has an encompassing definition best described as an ecosystem (Holling, 2001).

Meta-theory builds from the collective of other unit theories and is concerned more with the larger social context of a community of theories (Ritzer, 1988). Meta-theories transcend other theories and methods by referring to the sensemaking of unit theories and methods while deriving the contextual meaning of the unit theories and the data it draws (Van Gigch & Le Moigne, 1989; Wagner & Berger, 1985). The benefit of a meta-theory is that it provides a holistic orientation and perspective from a mix of isolated theoretical positions.

To advance the theoretic building blocks and synthesize the theories' utility through the application of a new ontology-driven conceptual modeling (Guizzardi, 2012), I draw attention to Figure 1 of the theoretical framework that shows the integrative meta-theory of city-level sustainable development approaches and the flourishing transformation of the city ecosystem. I define the city as a sustainable ecosystem as Figure 1 reveals the meta-integrated theoretical framework for sustainable development in cities as sustainable ecosystems (CASE) and the supporting theories that help inform the organizational structure and configurational processes of sustainable development.

9

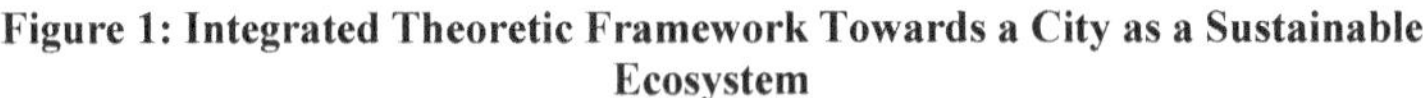

**Figure 1: Integrated Theoretic Framework Towards a City as a Sustainable Ecosystem**

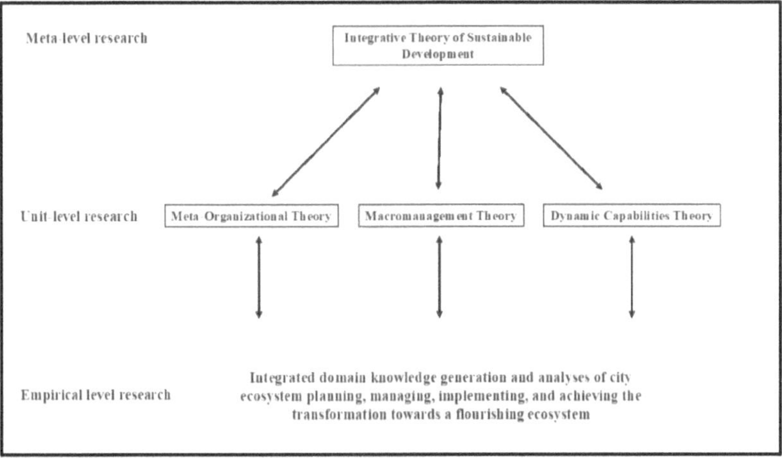


Figure 1 shows the relationship between meta-theories and theories. The meta-theoretical domain (meta-level) connects and informs between theories (unit-level) and the (empirical-level) knowledge generation, analyses, and integration. This integrative meta-theory for city-level sustainable development links the theoretical approaches of holistic organizing principles, positive action through macromanagement, and whole systems dynamic capabilities to develop an overarching framework that accommodates the wide range of "endogenous elements" of theories and their relationships as building blocks (Klein, Tosi, & Cannella, 1999). As the building blocks, the unit theories are referred to as conceptual lenses; and together with their inter-relationships, they form the infrastructure of the meta-theory (Ritzer, 1988). Since the lenses are developed at the meta-theoretical level, they are able to resolve fundamental paradoxes that exist within city ecosystem sustainable development and the underpinning unit theories. The primary

10

paradox I focus on is between the whole and its parts, in which I acknowledge and study the presence of both in this book. Although this paradox of external-internal ecosystems dynamics is my focus, I highlight other paradoxes that are underpinned and embedded within the ecosystem dynamic. For example, many theories of organizational capacity building are based on the concept of growth (Lester, 2002). Yet, in many ways, it is the dominant ethos of "economic growth over all" that is associated with unsustainable organizational routines and resource deterioration (Bayulken & Huisingh, 2015; Choi & Pattent, 2001). Meadows (1999) referred to rapid urbanization growth (e.g., population, economic, technological progress) as the origin of most problems such as social inequity, environmental destruction, resource depletion, and urban deterioration, which recognized the growth paradox as both the solution and the problem in development. However, the growth-sustainability paradox can be resolved at a dialectic level that acknowledges and encapsulates both development and stability (Stacey & Griffin, 2005), as well as circularity (Corona, Shen, Reike, Carreón, & Worrell, 2019) and decoupling (Jackson, 2016). The meta-theoretical integrative lens develops a resolution of the "whole and its parts" paradox, particularly in Holling's (2001) ecosystem perspective. Through the lens of an ecosystem, a holistic view of sustainable development is acknowledged by the relationships between the ecosystem implementing organization, change management approaches for growing and maintaining the ecosystem, the meta-organizational structures nested within the system for polycentric governance, the employment of dynamic capabilities for implementing and achieving the city ecosystem developmental goal of a flourishing city as a sustainable ecosystem.

To employ my own empirical evidence that validates and confirmed integrative ecosystem management as a transformational change approach, I framed the book with the meta-theory of ecosystem sustainable development. This meta-theoretic framing enabled us to employ a mixed-methods approach for the dissertation's research design, where each of my three studies was also theoretically framed using one or more of the unit theories. And with the combination of qualitative and quantitative methods, I triangulate the meta-theory of ecosystem sustainable development and provide the empirical evidence to validate and advance the theories, which additionally answers my primary research question. For example, in Study 1, I used grounded theory methodology to explore the factors practitioners and technical experts said were important for successful sustainable development implementation, and one factor that emerged was *meta-organization*. Study 2 was framed with *macromanagement, meta-organization*, and the *dynamic capabilities* theoretic frameworks. Study 3 was framed with *meta-organization* and the asset orchestration *dynamic capabilities* theoretic frameworks. Table 1 displays the study, the unit theory coverage, and the aspect of ecosystem sustainable development I examine.

**Table 1: Unit Theories, Study, and Empirical Aspect**

| | Integrated Meta-theory of Ecosystem Sustainable Development (ESD) Unit Theory Development | | | |
| --- | --- | --- | --- | --- |
| | Macromanagement | Meta-Organization | Dynamic Capabilities | Empirical Aspect of ESD |
| Study 1 | | X | | Success factors of ESD implementation |
| Study 2 | X | X | X | Integrated ecosystem management operationalized towards building ecosystem innovation and resilience |
| Study 3 | | X | X | City ecosystem performance |

As Table 1 shows, Study 1 examines factors that led to successful sustainable

development implementation in cities, according to practitioners and technical experts in

sustainable development. In Study 2, I examine Appreciative Inquiry operationalized to

validate its positive impact on ecosystem vitality by building innovative capacity and

resilience. Study 3 examines ecosystem performance and the combination of internal

factors that predict either high or low outcomes. Each of the different facets of ecosystem

performance, ecosystem vitality, and the sustainable development success factors will

contribute empirical evidence towards encapsulating the unit theories and meta-theory of

this book.

What I hope to accomplish with this theoretical approach is to advance integrative

ecosystem sustainable development theory and validate my conclusion that the

implementing meta-organization and community stakeholders are the internal drivers that

determine the sustainable development identity of the ecosystem. The evidence from my

collective studies will establish this by identifying the factors that determine successful

sustainable development planning, strategizing, and implementation, by conceptualizing

and operationalizing Appreciative Inquiry meta-organizing platforms for building

sustainable innovative capacity and resilience, and by highlighting the importance of meta-organization for developing dynamic capabilities, specifically, powerful leadership asset orchestration, which are two of a few factors that influence high city ecosystem performance. The integrated findings will identify the components (internal) and developmental stage (external) and also demonstrate how a city's implementing organization can plan, strategize, and design their city as a flourishing and sustainable ecosystem. They will explain the integrative meta-theory of city ecosystem sustainable development and elucidate how the integration of systems and resources by macromanagement, meta-organization, and dynamic capabilities functions as the primary driver for growth, maintenance, building resiliency, and further reorganizing systems and resources for building innovative capacity within the city ecosystem. Moreover, the integration of my findings will support my claim that the integration of macromanagement, meta-organization, and dynamic capabilities, or the integrative ecosystem management approach, is effective for city ecosystems' achievement of sustainable city goals. In the following section, the fundamental conceptual lenses for understanding integrative sustainable development and whole systems management approaches supported by city ecosystem planning, managing, implementing, and achieving the transformation towards a flourishing ecosystem.

# CHAPTER 3: LITERATURE REVIEW

For this literature review, I included the following databases in my search: EBSCO, Google Scholar, and Clarivate Web of Science. The first step of my review consisted of a search based on a structured key-word of these databases, using the following terms: "sustainable development," "flourishing," "ecosystems," dynamic capabilities," "systems approach," "macromanagement," "sustainable innovation," "Appreciative Inquiry," "cities," "community development," and "meta-organization." The objective of this literature review was to rigorously synthesize, review, and integrate representative literature on the topic of sustainable cities; and to develop an overarching framework that accommodates the wide range of "endogenous elements" of theories and their relationships as building blocks (Klein et al., 1999). The building blocks here are the unit theories referred to as conceptual lenses, and together with their inter-relationships, they form the infrastructure of the meta-theory (Ritzer, 1988). This meta-theoretical approach to my literature review builds and presents a framework on cities as sustainable ecosystem sustainable development (the external form, identity, and condition), structured from the ecosystem meta-organizing, dynamic capabilities, and macromanagement (the panarchic, internal systems factors, and dynamics).

Following my search, I collected, organized, and analyzed my research articles in NVivo. Two research articles emerged as having key theories to develop my meta-theoretic conceptual lens: "Understanding the complexity of economic, ecological, and social systems" (Holling, 2001) and "An integrative metatheory for organizational learning and sustainability in turbulent times" (Edwards, 2009). Supporting the unit level conceptual lens, I found the research articles: "Management, humanism, and society: The case for macromanagement theory" (McFarland, 1977); "The positive arc of systemic strengths: How appreciative inquiry and sustainable designing can bring out the best in human systems" (Cooperrider & McQuaid, 2012); "Meta-organizing firms' capabilities for sustainable innovation: A conceptual framework" (Berkowitz, 2018); "Understanding the elusive black box of dynamic capabilities" (Pavlou & El Sawy, 2011); and "Universities and innovation ecosystems: a dynamic capabilities perspective" (Heaton et al., 2019). Other articles concurred and supported attributes of each theory from these articles.

I focused my analyses on the meta-theory ecosystem sustainable development and the unit theories of macromanagement, meta-organization, and dynamic capabilities. I concentrated on integrative theory in my meta-theory and unit theories for a systematic synthesis as a whole meta-theory and theoretical framework. Ultimately, my goal was to synthesize the conceptual lenses to frame a meta-theory of sustainable ecosystem development by the meta-organizational governance structure, macromanagement systems approach, and dynamic capabilities. These conceptual lenses will provide insight into the whole systems integrative planning, managing, and implementing sustainable development goals in cities.

To approach the integrative meta-theory, I follow a stepwise literature review process. I first describe the elements of the ecosystem and development relevant to analyzing external and internal dynamics as it relates to city-level sustainable development. From my unit theories, I then build conceptual lenses that identify and describe certain characteristics and attributes of my unit theories that influence systems-level sustainable innovation and transform the city into a sustainable ecosystem. The third step of the review deepens my synthesis and features the integrated attributes of these theoretic lenses to illustrate how these attributes foster sustainable innovation by a macromanaging platform and process along with the development dynamic capabilities. I highlight the dynamic capabilities of sensing the environment, adaptive learning, coordination, and integration that help reconfigure and transform existing ecosystem organizational capabilities and into new ones that match a flourishing CASE.

**The Meta-Integrative Focal Lens to Cities as Sustainable Ecosystems**

To meet the challenges of sustainable development on the scale of cities involves deep structural transformations. The radical transformation needed for transitioning cities as sustainable ecosystems (CASE) while achieving Sustainable Development Goals (SDGs) involves paradigmatic shifts not only in the natural and artificial environments and structures, but also in how stakeholders in cities internalize values, purposes, and visions for future states of well-being and prosperity. As explained by Sachs, Schmidt-Traub, Mazzucato, Messner, Nakicenovic, and Rockström (2019), "...it will require complementary actions by governments, civil society, science, and business [for achieving SDGs]" (p. 805).

A benefit of the integrative conceptualized theoretical lens is that it provides stakeholders with a holistic view to understanding and approaching the whole ecosystem. In this case, the whole is represented as a city or economic region. Integrative theory allows the assessment of information about the internal factors and external influences, as well as the external factors and internal influences that interact to determine this mirroring systemic sustainability of both humans and the ecosystems they are settled (Holling, 2001). Moreover, the integrative framework emphasizes the interconnected nature of diverse elements and the multitude of dynamic interactions within the ecosystem. I find it more useful to source my theoretic domains with the combination of thought-provoking metaphors to think about human and ecosystem interactions and produce coherent patterns of structure, functions, and processes of highly complex systems nested within higher-ordered complex systems that I allegorize as ecosystems (Stacey, 2002).

Although there is seemingly an infinite number of heterogenous purposes, values, institutional logics, cultures, structural designs, business goals, and human intellect that are structured and interconnected within city ecosystems, there is a pattern of similarities inherent in these complex adaptive systems (CAS) that contributes to the ecosystem form. Holling (2001) explains the ecosystem form through four functional phases and the process functionality of "living" or "vitality" CAS. According to Holling (2001), ecosystems are complex adaptive and responsive systems with embedded communities and elements of living and nonliving systems that interact cyclically within the environments or within the boundaries of a confined morphogenic field.

In this case, the city ecosystem is the morphogenic field and model of a complex adaptive system (CAS). City ecosystems function through phases of birth, growth, maintenance, deterioration, and again to a rebirth continuing the ecosystem's vitality. What determines the phase of an ecosystem depends on the collective functioning of the smaller panarchic systems in the development of the ecosystem. For example, once a city emerges, and the community grows as an ecosystem, implementing organizations configure and link social, environmental, and economic systems for exploiting and conserving resources. As the second law of thermodynamics illustrates, this also contributes to the deterioration of resources such as the natural, built environments, and socially cohesive communities within the production and consumption sides of the economic system. However, the focus of this book is mostly on how these implementing organizations reorganize city ecosystem resources towards efficiency and beyond toward sustainable and flourishing developmental stages. Figure 2 illustrates the complex adaptive cycle of an ecosystem.

**Figure 2: Complex Adaptive Cycle of Ecosystems Model**

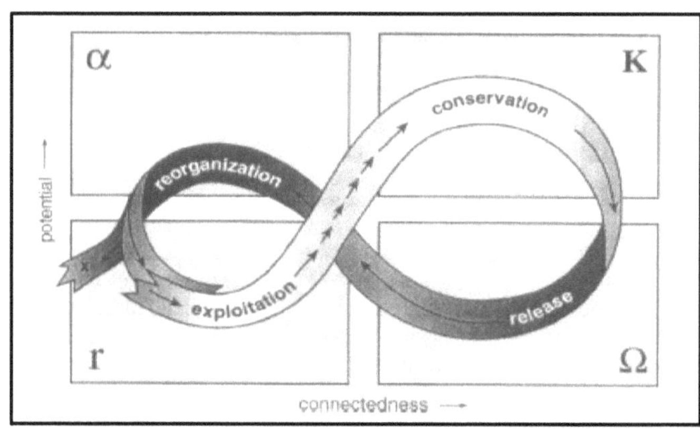


There are three properties in quadratic phases that shape the adaptive cycle and the future state within the boundaries of any particular ecosystem. The three properties are the *inherent potential* of the system (x gradient), which determines the range of future innovative possibilities; the *internal controllability* and degree of connectedness (y-gradient), which determines autonomy and flexibility of control; and *adaptive capacity*; which is the resilience of the system. The four functional quadrants represent α- (renewal/birth), r- (accumulation/coupling), K – (conserving/maturing), and Ω (opportunity/uncertainty). The infinite shaped flows between the four stages alternate between long periods of slow accumulation and transformation of resources (from exploitation to conservation, or r [accumulating/coupling] to K [conserving/maturing]), with shorter periods that create opportunities for innovation (from release to reorganization, or Ω [opportunity/uncertainty to a [renewal/birth]). Ecosystems are structured to function through these phases and process flows.

If we equate these functions to CASE, cities flow through similar phases to maintain their state of "living." For example, humans build, settle, and maintain infrastructures as well as cultural, economic, and social systems in the *birth* phase of a new city. The CASE grows by creating affinities that attract new human resources to cultivate (*exploit*), accumulate, and maintain vitality as the city grows toward maturity. Many cities in the developed and developing economies fall within the stage of *maturity*, since many of these global cities have existed for centuries. During the maturity phase, there are routine and dynamic interactions, that through the law of thermodynamics, create outputs of loss (*release*), deterioration, and waste from all of the accumulations artificially materialized by human interactions. For example, we construct buildings for living and working, water sewage, and electric systems that, over time, physically deteriorate. During the release phase, there is a higher degree of uncertainty of the future, as the ecosystem can fail, collapse, or totally deteriorate into a non-existence state unless the collective human interactions are guided to search for opportunities to adapt and reorganize for change. As humans recognize this release deterioration phase, by decision and choice, we sense and seize (*opportunities)* to develop ideas, strategies, and implementations to upgrade and reconfigure (*reorganize*) the city system assets and infrastructure into new possibilities to materialize the vitality of the ecosystem. The assumed goals of CASE are growing the internal essential system assets by human interactions while maintaining the overall structure and identity of the CASE towards a new *rebirth* of the city. As it relates to the achievement of SDGs, this city rebirth has been translated as smart cities, sustainable cities, green cities, and flourishing cities. Just as the ecosystem's renewal phase continues "living" by the functional process of cultivating, conserving, releasing, and reorganizing, the CASE expresses the "living-system" analogy by its adaptive capacity and resilience through the complementary social designing and asset orchestration of the nested panarchic systems of human interactions towards a healthy "living" ecosystem vitality (Luther & Flora, n.d.). Scholars referred to this polycentric collective action and asset orchestration by managers, leaders, and influencers of cities as "the decision system," an intermediate panarchic system nested within the city (Homsy & Warner, 2015).

In contrast, a critique of assessing cities as living ecosystems would be challenged by Maturana in his seminal work of autopoiesis and living systems. Maturana (1981) posited that a distinguishing characteristic between allopoietic and autopoietic systems is that autopoietic systems have regenerative capabilities, creating manifestations of themselves. Since cities do not create other cities, Maturana challenges my notion of CASE. However, I argue that cities are more than just geographically located physical artifacts designed by humans. Through the integrative lens, I argue that the scope of understanding the city as an ecosystem is a broader perspective because it acknowledges not only the artificial elements of the city such as the built structural components of a city, but more importantly, they also include the human, social, cultural, political, economic, and natural elements as well. It is the combination of human factors and how we reconfigure the elements that establishes the emergence of cities. Furthermore, if cities are more than built environments, and humans are living systems; the mirroring effects between collect human activities and city ecosystems reveals the gestalt of a city, since they are one and the same within the morphogenic field.

In contrast to the growth/innovation, the adaptive capacity/resiliency in CASE is another characteristic integral to most living systems and is best described as the maintenance property of the system. As it is related to sustainable development in cities, resiliency is aligned with cultivating community capabilities and how city stakeholders respond by discovering opportunities through collective strengths to cope with external stresses and disturbances because of social, political, economic, cultural, and environmental change (Bec et al., 2019; Dale et al., 2010). The reflexivity between the emergence of innovation capacity and resilience of a city ecosystem functions as both a response mechanism to internal and external perturbation and a functional mechanism for creative destruction and reorganization opportunities.

Between human systems and their CASE designing are intermediate systems such as social networks, organizations, and institutions that shape and give the city an identity and form. Through the decision systems of city mayors, executive directors, politicians, business leaders, municipal workers, faith-based leaders, non-profit managers, community activists, socio-cultural and socio-technical designers, and all other informal leadership and peripheral stakeholders contribute to the CASE. Whether working by disintegrated or interconnected approaches, the collective actions of leaders, managers, developers, social and cultural influencers having some degree of the ecosystems' asset distributive power and that guide city decision-making can be defined as the meta-organizational structure architects, macro-managers, and components of the decision system (Berkowitz & Dumez, 2016; Cooperrider & McQuaid, 2012; Gulati, Puranam, & Tushman, 2012; Holling, 2001).  These organizing structures function through phases mostly flowing through phases at rates slower than individual humans but at faster rates than the larger hierarchal systems such as institutions and CASE (Holling, 2001). Each lower-level panarchic system has an influence on the higher-ordered panarchic systems, which also relates to Herbert Simon's (2019) hierarchies and semi-autonomous levels form higher-ordered structures from interactions among other systems of similar speed. Figure 3 illustrates the order of nested panarchic systems and how they are connected and influenced by each other.

**Figure 3: Hierarchal Order of Panarchic Complex Systems on Ecosystems**

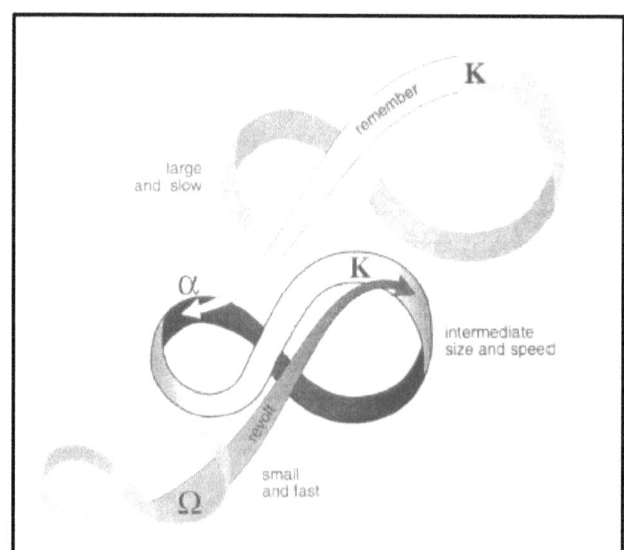

Starting with the small hierarchal system at the bottom in Figure 3, this represents

what can be interpreted as a human system, stakeholder, or person. The intermediate-

sized system can be interpreted as an innovation ecosystem or the organizing of

stakeholders and organizations for fostering innovation and resilience within a city

ecosystem. The large hierarchal system can be interpreted as the city ecosystem. Because

of the size, each nested hierarchal system has a variable speed of reaction, reflexivity,

adaptivity, and reconfiguration capabilities that is faster if smaller and slower if larger

(Holling, 2001). For example, people can move, think, change, react, revolt, and

collaborate faster individually compared to an organization. Moreover, an organization

can reconfigure for competitive advantage faster than institutional change, which can be

faster than a city ecosystem transformation.

The CASE perspective is an inclusive one that reveals the patterns of humans as

parts of local socio-cultural economic and socio-ecological systems. These patterns of

human routines and dynamic interactions are recognized within the conditional

boundaries of specific bioregions to the biosphere, in which the focus is on the

humanistic relationships, social designing, and processes that support life in its myriad of

complex forms in partnerships and cooperation as well as competition and conflicts

(Newman & Jennings, 2012).

The concept of CASE ties humanistic interactions firmly within the ecosystem

and not apart from it. Humans and natural ecological systems are part of the whole

systems, in which we are influenced by the natural ecological (living) and artificial

(human-created) subsystems within which societies and nature function. As it relates to

the management of achieving SDG goals, the CASE model provides the lens for multi-stakeholders to view cities and regional economies as open systems that comprise integrated autopoietic and allopoietic systems properties. Describing cities as hybridized living ecosystems can provide consideration and helpful guidance for transforming human adaptive-responsive processes as well as the CASE adaptive-responsiveness while simultaneously designing, reconfiguring, and restoring natural and artificial environments in a harmonious congruence.

**The Developmental Focal Lens to Flourishing CASE**

Innovation for sustainable development has been widely viewed as the key to transitioning societies and their environments to sustainable and flourishing futures (Silvestre & Țîrcă, 2019). To achieve sustainable development success and the goal of sustainable cities, it will take an integrative ecosystem management process (Sroufe, 2018) of creating social innovation (Westley & McGowan, 2017), ecological modernization (Mol, 2000), the mobilization of sustainable education, political and institutional responses by policies, and civic participation (Sachs et al., 2019). An integrative ecosystem management approach is needed for growing sustainable city ecosystem infrastructure, industry, and innovation, while simultaneously improving social arrangements for upward mobility. In generic terms, an integrative ecosystem management approach develops adaptive capacities for maintaining city ecosystems for resilience as a response to climate change, as well as ecological and social perturbation.

As a collection of systemic "wicked" problems, the challenges of achieving sustainable cities are multidimensional; but developments in practice and empirical research conducted over the last five decades have produced fundamental methods of addressing these wicked problems on an individual basis. However, over the last decade, empirical research and practical, sustainable development have emerged on approaching and evaluating sustainability broadly in more comprehensive and holistic ways (Bai, Surveyer, Elmqvist, Gatzweiler, Güneralp, Parnell et al., 2016). Yet, scholars and practitioners have not agreed upon sustainable development characterization and measurement such as what is to be sustained, what is to be developed, the linkage between social and environmental development, and for what extent of time (Parris & Kates, 2003). However, the United Nations 17 SDGs were designed by academics and practitioners in the global community, and over the last five years, these goals have been adopted by leaders in cities, communities, and organizations as an evaluative tool for localizing sustainable development. With over 160 indicators embedded within 16 of the goals, each indicator is quantified by a score. These scores translate into improvements and declinations of a city's specific sustainability indicator, and the compilation of scores results in a total score representing a quantitative level and comparative ranking of achieving the 17 SDGs. The 17 SDGs can be a comprehensive guide for practitioners and communities to evaluate their sustainable development progress; however, it argued that the score translation is linear and doesn't capture internal human transformation indicators necessary for assessing and evaluating holistic developmental changes within the sustainable developmental lens of a city ecosystem (Sachs et al., 2019). However, the 17 SDGs can complement the humanistic and city ecosystem developmental lenses of transformation as a unit (Rant, 2020).

Developmental changes on the ecosystem level require a degree of understanding complexity in transformation because of the multi-levelness and non-linear nature of complex systems (Cacioppe & Edwards, 2005; Edwards, 2005). Systems-level developmental change can be viewed by movement through stages. Organizational

systems such as cities progress or regress through stages. The progression of stages

depends on various factors, approaches, social structures, and guidance pursued by

leaders and stakeholders within cities. Although there are guidelines for sustainable

development established by the United Nations 17 SDGs, scholars have found that

stakeholders in cities lack the awareness that the SDGs exist and lack an understanding of

how the 17 SDGs can be operationalized (Sachs et al., 2019). In addition, there is still

confusion and significant conscious resistance to sustainable development, where

antagonizing agents either do not believe in the institutional logics of sustainability,

doubt how the SDGs translate into system transformation or do not have the agency to

change from their normative behavior and routines. Scholars have found this resistance to

be apparent across many social class topographies and demographics (Bain,

Kroonenberg, Johansson, Milfont, Crimston, Kurz et al., 2019; Linnér & Wibeck, 2020).

It also has been empirically studied and revealed that traditional socio-economic and

socio-ecological system models, behavioral interventions, independent organizational

effectiveness, and change management approaches to sustainability based on only

reductivism have not worked in favor of systems as a whole, nor have been effective of

whole systems social transformation (Cooperrider & McQuaid, 2012; McQuaid, 2019).

Aggregated approaches for behavioral change implemented by organizational change

agents, non-profit organizations, and government leaders have yielded incremental

progress towards sustainable development (McQuaid, 2019). Dominant top-down

technocratic management approaches and policymaking have not been effectively implemented nor have been found to fully transition the social arrangement in cities

towards sustainability (Saha & Paterson, 2008).

Although it is difficult to assess and view developmental progression and success

as a whole, a stage-based model has been developed for organizational sustainability

(Edwards, 2005, 2009). The stage-based model was the result of empirical work

combined by meta-theoretical combinations in the developmental lens (Dunfy, Griffiths,

& Benn, 2007; Laszlo & Brown, 2014; Pavez, Kendall, & Laszlo, 2020; Van Marrewijk

& Becker, 2004; Van Marrewijk & Hardjono, 2003; Van Marrewijk & Werre, 2003). The

various stages represent qualitatively different design archetypes that are associated with

multiple levels of organizational sustainability. Even though the majority of the empirical

studies of this model were analyzed on the firm level, I adopt this stage-based model on

the city ecosystem level. I present the stage-based model because it provides a

developmental lens for sustainability identifying the qualitative spectrum of

transformations that are potentially available to city ecosystems. Each organizational

ecosystem form is associated with various internal qualities and exterior conditions that

are interconnected, as explained by the integrative human-ecosystem qualities of complex

adaptive systems. The systems' forms are as follows:[1]

- *Subsistence ecosystem organization.* Sustainability is observed as a survival
  mode. The values base is one of working hard and getting by without doing
  obvious damage or reducing damage to individuals or environments. Survival
  and maximization of profit are regarded as the sole organizing principle and
  purpose of business industries with social welfare safeguards for the disabled,
  elderly, and children. The social arrangement is modulated by homophilic
  networks separated by class and status and unequal community amenities.

---

[1] Adapted from Edwards, M. G. (2009). An integrative metatheory for organisational learning and
sustainability in turbulent times. The learning organization.

- *Avoidant ecosystem organization.* Sustainability is observed as a revolt against dominant social norms by oppositional grassroots groups, communities, and emerging institutional logics. There is a general ignorance of ethical standards and legal responsibilities in business and social communities. Environmental damage is ignored as negative externalities at the expense of "doing business as usual." Disinterest is the prevailing attitude towards the impact of systematic activities on the workforce in local industries and communities. The local government and business communities operate mostly from top-down hierarchal intervention to any change mechanisms or policies.

- *Compliant ecosystem organization.* Sustainability is regarded as an impost. The compliant systems organization supports industry regulation as a way of circumventing more demanding regulations regarding sustainability. The business communities reactively respond only to state and federal regulatory requirements and policies as they are applied. Grass roots movements for sustainable change utilize social mobilizing structures to voice concerns for disenfranchised and deteriorating communities and to stimulate policy reform toward government and institutional leadership.

- *Efficient ecosystem organization.* Sustainability is valued as a source of cost saving (i.e., the "business case" for sustainable development within the economic systems and industries) and as a source for social equity programming within the community. This stage is observed as optional activities that attempt the triple bottom line protocol in business, with limited local policy programming. Sustainability is defined in terms of helping organizations to continue business in changing social environments adopting a business purposes that focus more on stakeholders, but not at the expense of profit.

- *Committed ecosystem organization.* The organization is committed in principle to economic, environmental, and social sustainability and goes beyond legal compliance. Local governments and non-profit organizations (NGOs) are also committed to improving social arrangement and social mobilization of marginalized groups in the community.

- *Sustaining ecosystem organization.* Sustainability is valued as a way of developing business organizations and industries, as well as multiple stakeholders on all fronts. Governments, business communities, and civic participants approach sustainable development by innovation strategies and inclusiveness of upward social mobility of all communities and stakeholders within the ecosystem. Transformational strategies are enacted for moving the city ecosystem towards sustainable development goal achievements and benefiting all communities within the ecosystem.

- *Flourishing ecosystem organization.* Sustainability is embedded within all aspects of the organizational system and is observed in holistic and intergenerational terms. Promotes a consciousness of flourishing on all social, environmental, and economic levels by actively creating sustainable innovation, sustainable communities and infrastructure, and sustainable enterprises. Sustainability refers to numerous layers of purpose, including avoiding harm to people and the natural environment, social well-being, economic prosperity, along with human emotional, social, and consciousness of spiritual/deep meaning and connectedness. (Adapted from Edwards, 2009)

As each developmental quality stage is explained, a greater degree of complexity and new paradigms of sustainability emerge, where new core capacities of the organizational system build inclusively from capacities of the previous stages. Figure 4 illustrates this inclusive nature and movement through the stage gradient. The developmental lens is structured accordingly from pre-conventional to post-post conventional stages, where the progressive inclusion of the inner stage forms the outer stages. However, there can be multiple pathways to sustainable development goals implemented by city leaders and stakeholders, and it may not always be a linear progression (or regression) from one stage to the next. City ecosystems can lose systems adaptive capacities and population, causing cities and their stakeholders to regress into worsened or unsustainable stages of sustainability. Because of the increasingly competitive market, businesses and industries in a particular city ecosystem may also move away, deteriorate, or become obsolete, which can also have profound effects on the social environment and opportunities. These factors and many others form the ecosystem in a regressed stage. Lastly, as with most transformations, evidence has shown that true transformation, especially at the scale of a city ecosystem, will involve considerable organizational disruption and "pain;" but with good leadership, a shared vision, government intervention, attainable resources, and civic participation, cities can adopt an integrated management strategy to achieve the transformation aimed (Colombo &

Delmastro, 2002).

**Figure 4: Stages of Sustainability in Ecosystems**

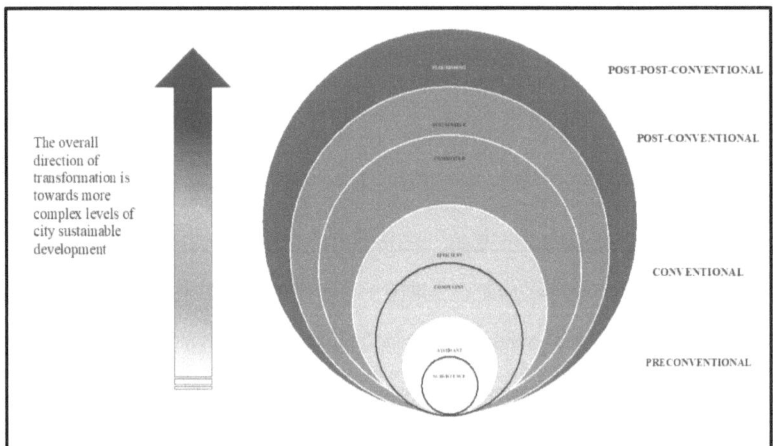


City ecosystems with plans and approaches at the post-conventional stage, which

are committed to adopting principles of sustainable development practices within the

culture and systems, will also retain the positive capacity to function and maintain at

conventional stages of system "efficiency" and "compliance." The conventional stages, in

turn, include capacities of surviving and maintaining institutional requirements for

governing and managing the ecosystem as a competitive and attractive city. Not too many

want to live in a city with constant chaos and very little governance for social order and

few amenities. Stage-based capacity building supports the internal stage, and the next

stage outward includes a greater capacity for engaging complexities and integrating

larger swaths of environment and social systems. The more ambitious the sustainable

31

development goal for the city ecosystem, the more complex the organizational system culture and structure needed to achieve sustainable and flourishing city ecosystems (Van Marrewijk & Werre, 2003). The higher aims of flourishing-as-sustainable require more expansive forms of sustainable development and meta-organizational integration.

Currently, there is no unilateral approach for broadly achieving a sustainable city nor securing human well-being on a mass scale. Strong evidence suggests that the advancement of an integrating systems approach for planning, developing, and managing towards achieving flourishing sustainable cities presents as an emerging integration strategy that may benefit the organizing, planning, and implementation towards transitioning city ecosystems (Yigitcanlar & Kamruzzaman, 2015). There is empirical evidence of meta-organizational attributes as a device for ecosystem governance, dynamic capabilities that ecosystems need to acquire to develop sustainable innovation, along with a macromanagement strategy that has relevance to achieving whole systems transformation. These theories as a conceptual lens will be described in detail, following a synthesis that features how attributes of this lens can foster sustainable innovation by macromanagement integration towards CASE.

**The Macromanagement Lens for Whole Systems Flourishing**

The macromanagement framework is useful in sustainable development in city ecosystems because it focuses on the interinstitutional relationships of social, cultural, ecological, and political values. Within the literature streams, an agreed-upon definition of macromanagement is conceptually still in development; however, there are some foundational elements and meanings of macromanagement I find relevant to city ecosystem development and innovation. One meaning of macromanagement is understood in contrast to the closely observable and controlling managerial approaches and aligns more with "stepping back" and managing from afar. This meaning can be interpreted as managers taking a holistic view to strategic managerial approaches. Some scholars have defined macromanagement as social organizations with an integrative approach to management that have developed into social institutions, whose purpose and goals are to serve society and plan for the future while balancing and influencing the social arrangement in the present (Gulati, Mayo, & Nohria, 2016; McFarland, 1977). Moreover, these social institutions can be understood as "the patterns of rules, customs, norms, beliefs, and roles which are instrumentally related to the needs of and purposes of society..." (Wells, 1970: 3-11), which represent "..the linkage of social structures of government, economies, businesses, education, and religious systems (McFarland, 1977: 614).

In earlier development of macromanagement, McFarland (1977) highlighted the controversies and criticisms of this social pluralistic lens and approach by focusing on the hierarchal conflicts embedded in managing social structures, "...with power struggles among contending elites and the interplay of autonomy, interdependence, and prestige among various organizations of the institutional matrix" (p. 614). But, as macromanagement theory developed, evidence emerged within the social sciences, public administration, and management literature streams revealing that social crises and stresses were unlikely to be solved by the dominance of any one single social institution but would require the fusion of interinstitutional management processes to influence social behavior patterns (Clarke, 1987; Kemp & Rotmans, 2005; McFarland, 1977). Some scholars have criticized macromanagement for sustainable development as being too utopian since cities, and nations for that matter, naturally operate unsustainably (Morley, n.d.; Oliveira, Andrade, & Makse, 2014; Rousseau, n.d.; Varro, n.d.). Furthering the argument, some scholars have illuminated the problems associated with sustainable development macromanagement in the forming of alliances within competing industries and institutions as well as the difficulties of motivating civil participation; especially a whole systems integrated management approach in the planning, managing and implementing processes of sustainable development (Olawumi, Chan, Wong, & Chan, 2018; Portney, 2011). For macromanagement of ecosystem sustainable development, critics have also found public views of sustainability and sustainable development are mixed; ascertaining the difficulties of coming to a consensus of competing institutional logics (Bain et al., 2019). Even more, other scholars and practitioners have voiced concerns about the scale of city ecosystem macromanagement efforts (Yigitcanlar, 2010), while others voiced concerns about measurability of subjective sustainable cities outcomes (Böhringer & Jochem, 2007) and how far in the future should planning efforts be set (Fuller, 2008; Lindblom, 1999).

In contrast to macromanagement critics, more recent studies concur with the notion of interinstitutional action by integrating the implementation stakeholder process (Weymouth & Hartz-Karp, 2018), integrating top-down and participative management (Dunphy, 2000; Edwards, 2005), localizing SDG achievement (Rousseau, Berrone, & Gelabert, 2019), advancing a systems approach (Bai et al., 2016), and linking horizontal and vertical socio-technical systems (Bai, Wieczorek, Kaneko, Lisson, & Contreras, 2009). Macromanagement has emerged as a viable approach to city ecosystem sustainable development because isolated reductionist approaches to social problems, the function of government alone, and the function of business alone are not singularly

sufficient to solve the wicked problems associated with cities and societies within them

(Cooperrider & McQuaid, 2012; Rajabi & Hesarinejad, 2013; Sachs et al., 2019;

Salarpour, Amiri, & Mousavi, 2019; Whitney, Trosten-Bloom, & Vianello, 2019).

However, by muddling through the foundational concepts of macromanagement, I found

some commonalities by definition to summarize a meaning of a macromanagement

approach that fits my narrative. My meaning of macromanagement is one that

consolidates heterogenous interinstitutional actions within socially pluralistic

arrangements seeking to influence present and future social arrangements for good

through holistic and deliberative democratic conversations and interactions (Cooperrider

& McQuaid, 2012; McFarland, 1974; Weymouth & Hartz-Karp, 2018).

Beyond the property of integrating collective action, there are other distinct

properties of macromanagement that attribute to its uniqueness compared to other

organizational systems approaches. First, embedded in macromanagement is a positive

focus on the "whole," along with generative employments of conversations and ideas. For

example, during the planning and strategizing for system change, the knowledge sharing

of information and idea generation would have a positive perspective, design, and

conversation. This means instead of strategic planning from problems and negative

consequences that emerge, then approach them with reductionist strategies; conversations

in strategic planning would focus on positivity and invert the contextual meaning of a

particular problem into an opportunity to innovate positive change (Fredrickson &

Losada, 2005; Roepke & Seligman, 2015).

A second property of macromanagement is the element of prospection. Prospection is the capability of being future-focused driven by optimism. Combined with a positive outlook, prospection serves as a mechanism for leaders, managers, and other stakeholders to think by optimism and forwardly productive management; and this way of thinking and doing exerts an optimistic future-focused controllability within social environments (Na, Chung, Jung, Hula, Fiore, Dayan et al., 2019). In contrast, backward thinking and management with pessimistic outlooks of social environments have languished stakeholders' controllability and capabilities associated with negative effects and senses of hopelessness and helplessness in social environments (Maier & Seligman, 2016; Seligman, Railton, Baumeister, & Sripada, 2013). This insight into prospection has been studied in social environments, such as city ecosystems, and has been found to be an instrumental social managerial capability when assessing and implementing public policy for socio-economic well-being indicators (Adler & Seligman, 2016; Roepke & Seligman, 2015). The macromanagement meta-epistemological elements of wholeness (vertical and horizontal polycentric action) with a positive future-focused approach and outlook is the essential management strategy for multi-stakeholder processes and integrative ecosystem management for the developmental goal of CASE.

As an instrumental multi-stakeholder strategy for fostering innovation for sustainable development in city ecosystems, I posit that the macromanagement approach synthesizes efforts for social benefit from collective industries with institutional and organizational leadership as well as grass-roots social movements and other informal social systems. Considered a third form of management, this approach includes the synthesis of top-down and bottom-up approaches (Cooperrider & McQuaid, 2012) and is by nature inclusive across the spatial social matrix of the ecosystem. The top-down management capabilities are considered the well-managed technical expertise, or hard power, while the bottom-up, or soft power, brings a distinctive strength of inspiration. Together, the combination of both strengths has the potential of creating exponential innovative power and capacity. I focus on the macromanagement approach within the innovation ecosystems because innovation is widely considered to hold the keys for whole systems building, such as city ecosystems (Berkowitz & Dumez, 2016; Boons & Lüdeke-Freund, 2013; Cassiolato, Pessoa de Matos, & Lastres, 2014; Clay, 2017; Henrekson, 2014; Macharis & Kin, 2017; Schaefer, Corner, & Kearins, 2015).

The macromanagement approach (MA) is not only a social construction of internal connectedness and controllability that promotes autonomy and empowerment within the ecosystem, but it also has an inherent potential for future possibilities towards flourishing by the adaptive capacities created by human ingenuity. Innovation emerges from conversations and the sharing of knowledge in human organizational systems that allows the process for future development and human abstractness to materialize that forms the ecosystem identity (Whitney, 2012). As stated by Hartmann (1991: 275) and Fry (2012: 54), "Words create worlds." This notion not only represents the innovative materialization of external built environments but more importantly, it symbolizes the human innovative process of internalizing a transformation and empowerment (Nikkhah & Redzuan, 2009) through the sharing of information and conversation (Berkowitz, 2018).

There is empirical evidence of employing MAs that encourages conversations and collaborative actions that enable knowledge transfer to foster innovation. Stakeholders were found to be empowered when performing interdependent tasks within communities of practice from core managers' advocacy of individual autonomy (Kirkman, Mathieu, Cordery, Rosen, & Kukenberger, 2011). Cozzens and Sutz (2014) and von Hippel (2009) was concluded that the merging of technical knowledge with local knowledge is an effective approach for local ecosystem innovation. The innovation ecosystem MA has been well established as an effective means for fostering sustainable innovation (Berkowitz, 2018) by systems-based innovation (Fuller, Bartl, Ernst, & Muhlbacher, 2004), and stakeholder innovation within multi-sector communities (Bunn, Savage, & Holloway, 2002). However, a macromanagement approach for the challenge of sustainable development that enables and stimulates the innovative capacity in the built ecosystem while empowering and shifting the social transformation towards a mass consciousness of flourishing is less studied.

In the management and positive action research literature streams, Appreciative Inquiry platforms have been well established as a device for macromanagement (Cooperrider, Whitney, & Stavros, 2003; Pandey & Gupta, 2008; Stavros & Saint, 2010; Whitney et al., 2019). There are a plethora of qualitative empirical studies examining Appreciative Inquiry AI and whole systems strengths-based strategies, such as the SOAR framework for macro-managers and stakeholder value creation and collaborative capacity building for organizational systems (Cooperrider, Godwin, Boland, & Avital, 2012) toward designing systems for flourishing (Cooperrider et al., 2003; Stavros, Torres, & Cooperrider, 2018). AI is the operating system of SOAR, while SOAR is process embedded in the system (Cole, Cox, & Stavros, 2019). SOAR is a positive-focused macromanagement approach to strategic thinking, planning, and leading to positive change for the whole (Stavros & Hinrichs, 2009). Both AI and SOAR engage multiple

stakeholders in generative, positive, and holistic change when applied to whole systems

change management. SOAR is a positive, generative, and strengths-based framework,

"…with a participatory approach to strategic analysis, strategy development, and

organizational change" (Stavros & Saint, 2010: x). AI is embedded into SOAR as a

strategic planning framework to create a transformational process that inspires

organizations, communities, and stakeholders to engage in results-oriented strategic

planning efforts (Cooperrider et al., 2003). In contrast to the strengths, weakness,

opportunity, threats (SWOT) strategic analytical tool, SOAR is a dialogical framework

that shifts stakeholders' conversations and focuses on imagining desired positive

outcomes and innovating to the aspired goal(s). Instead of a strategy that focuses on

problem-solving weaknesses and threats, SOAR enables the discovery of strengths for

stakeholders to transform a system towards continuous improvement.

**The Meta-organization Lens of Sustainable Innovation Governance**

A meta-organizational design structured in city ecosystems can be seen as an

organizational governance device for sustainable development and innovation. Meta-

organizations, simply defined by Gulati et al. (2012), are the organizing of organizations.

Meta-organizations can be made exclusively of businesses in an industry, but they can

also be mixed, as in organizations and individuals or in multistakeholder groups that

collaborate in civil society. The conceptual meaning meta-organizing describes stems

from older phenomena, such as was inter-organizational collaboration interest (Streeck &

Schmitter, 1985) and industry alliances for business (Van Waarden, 1992). As the

concept of meta-organizations evolved in the literature, it described a broader range of diverse structures encompassing formal (Ahrne & Brunsson, 2008) and informal (Gulati

& Puranam, 2009) organizations. For example, the growing body of literature includes

the meta-organization lens in digital ecosystem platforms (Gawer, 2014), sharing

economy platforms (Berkowitz & Souchaud, 2019), trade unions (Furaker, 2020), and

multistakeholder processes for sustainable development and innovation (Berkowitz,

2018; Berkowitz, Crowder, & Brooks, 2020; Chaudhury, Ventresca, Thornton, Helfgott,

Sova, Baral et al., 2016; Valente & Oliver, 2018). With the inclusion of political and

social institutions, business firms, and non-profit organization, meta-organization brought

forth the growing importance of collective action among organizations in empirical

research beyond the isolated boundaries of just a firm, industry, government, social

movement, or non-governmental organization. As Ahrne and Brunsson (2008) proposed,

meta-organizations function differently from organizations of individuals because they

rely more on member organizations' resources, lack hierarchy, and rely on the consensus

to make collective decisions.

Even though the knowledge generation of meta-organizations is growing, there is

a literature gap and opportunity to develop a meta-organizational framework for

sustainable development in city ecosystems. By applying this conceptual lens, I seek to

build upon the meta-organizational theoretical approach of sustainable development and

innovation governance. To do so, I integrate the meta-organization approach in a triple

lens: (1) to uncover attributes of meta-organization in city ecosystems as an instrument

for the governance of a shared vision goal; (2) to identify the dynamic capabilities city

stakeholders need to employ for sustainable innovation fostering; and (3) express how the

meta-organizational governance structure attributes relate to the ecosystem's sustainable innovation dynamic capabilities. The meta-organizing lens is particularly important to the phenomena of my empirical work because it assists the sustainable development and innovation narrative encompassing the broader view of underpinning city ecosystem organizational dynamics of decision-making and systematic organizational effectiveness towards sustainable cities. I draw specific attention to the dynamic capabilities concept of integrative reconfiguration, which has been somewhat overlooked within the literature.

Before we uncover attributes of meta-organizing and the proposed ecosystem governance consideration, we first must define sustainable innovation and how it has been governed and fostered in its development. There are many descriptive of sustainable innovation (SI), and scholars have yet to agree upon a single definition. The European Union follows the UN's definition of SI as artifacts "consisting of economic, social, and environmental benefits" (Von Schomberg, 2011: 9). Other scholars have described SI as an innovation that improves sustainability performance (Carrillo-Hermosilla, Del Río, & Könnölä, 2010) or an innovation that avoids harming people and the planet by offering new products, services, and technologies that foster sustainable development (Voegtlin & Scherer, 2017). I define SI with the common scholarly agreement that SI is one that aligns with social needs and sustainability impacts (De Saille, 2015), leading to socially desirable outcomes (Stahl, Eden, & Jirotka, 2013).

Literature in fostering the development and diffusion of sustainable innovation has revealed the discourse and opportunities in the governance framework. Reactive policy approaches of governance frameworks based on regulation have not been as effective as planned (Berkowitz, 2018; Lee & Petts, 2013). Empirical studies have revealed that little to no transformative power on norms, discourses, institutions remain vague, particularly in sustainable and responsible research and innovation frameworks (De Saille, 2015). Koops (2015) mentioned, "existing governance structures and regulatory frameworks can sometimes be poorly equipped to accommodate certain innovations or system transitions […] this underlines the importance of integrating the governance perspective within the whole cycle of responsible innovation processes, so that technologies, practices, and governance can fruitfully co-evolve" (p. 10). This notion prompted a research agenda to explore how tackling grand challenges require specific governance instruments, especially on the scale of ecosystem sustainable innovation. Scholars such as Davies and Horst (2015) and Cooperrider and McQuaid (2012) concur with this argument that a governance reconfiguration is needed for sustainable innovation and city-level sustainable development transformation. I, as well as these scholars, am not proposing a systematic social revolt in government, but a governance device that aims to shift some of the weight from top-down, hard power, regulatory frameworks to soft power, bottom-up frameworks with horizontal distributions of governance. Similarly, within the quality management (QM) literature streams, there is also evidence that a decentralized structure of hard QM combined with a strong integration between functions of soft QM was found favorable and beneficial for fostering product innovativeness (Zeng, Zhang, Matsui, & Zhao, 2017). In the context of city ecosystem innovation, this entails participatory governance with multi-stakeholder involvement and deliberative democracy based on meta-organization (Berkowitz, 2018; Fung & Wright, 2003; Weymouth & Hartz-Karp, 2018).

Meta-organizations have been seen as a post-conventional organizing principle (Edwards, 2009), and by some perspectives, unconventional organization (Bres, Raufflet,

& Boghossian, 2018). But meta-organization plays a key role in collectively building capacities and capabilities within the boundaries of an ecosystem, especially in the achievement of sustainable cities and social transformation for adopting sustainable practices and SDGs (Berkowitz, Bucheli, & Dumez, 2017; Sachs et al., 2019). As I will examine in this book, meta-organization can play a key role in building CASE adaptive capacities (Folke, Carpenter, Elmqvist, Gunderson, Holling, & Walker, 2002; Holling, 2001) and macromanagement dynamic capabilities (Heaton et al., 2019; Schoemaker, Heaton, & Teece, 2018; Teece, 2007; Teece & Leih, 2016; Teece, Pisano, & Shuen, 1997).

Meta-organization as a governance instrument for firms has been examined in the context of assisting the building and diffusion of organizational capabilities for sustainable innovations (Berkowitz, 2018). However, little work has examined the utility of this governance device on the building and diffusion of dynamic capabilities in city ecosystems, that is, integrated organizing and reconfiguring of adoption and practices at the ecosystem, organizational, institutional, and entrepreneur levels. First, we must identify the characteristics and attributes of meta-organizations and how they impact governance mechanisms. Table 2 synthesizes the attributes of meta-governance, that is, governance by meta-organization (Berkowitz, 2018: 423).

**Table 2: Attributes of Meta-Organization as a Device for Governance**

| Attributes of meta-organizations | Citation from Literature |
| --- | --- |
| Act as a platform | (Ciborra, 1996; Gawer & Cusumano, 2014) |
| Has a low-cost structure | (Berkowitz & Bor, 2018; Berkowitz & Dumez, 2016) |
| Consists of an association of other organizations | (Ahrne & Brunsson, 2008; Berkowitz & Bor, 2018) |
| Results in self-regulation and city ecosystem reconfiguration | (Heaton et al., 2019; Lawton, Rajwani, & Minto, 2018; Rajwani, Lawton, & Phillips, 2015; Spillman, 2018) |
| Makes decisions on consensus | (Ahrne & Brunsson, 2008; Bor, 2014) |
| Functions by heterarchy rather than hierarchy | (Ahrne & Brunsson, 2008; Stark, 2009) |
| Inclusive of heterogeneous actors from civic society, non-profits, firms, government officials, institutional members, and academia | (Berkowitz et al., 2017; Boström, 2006; Clarysse, Wright, Bruneel, & Mahajan, 2014; Sachs, Schmidt-Traub, Kroll, Lafortune, Fuller, & Woelm, 2020) |
| Produces reporting mechanisms and enhances social responsibility and sustainable development | (Berkowitz et al., 2017; Chaudhury et al., 2016; Hansson, Arfvidsson, & Simon, 2019; Rasche, Waddock, & McIntosh, 2013)) |
| Facilitates knowledge transfers and collective learning | (Ahrne & Brunsson, 2008; Clarysse et al., 2014; Gulati et al., 2012) |
| Permits coopetition, competing institutional logics, dialogue, and information capture | (Berkowitz et al., 2017; Cooperrider & McQuaid, 2012) |
| Produces information for members and outreaching | (Berkowitz & Bor, 2018; Malcourant, Vas, & Zintz, 2015) |

Attributes of meta-governance, that is, governance by meta-organization, were previously synthesized by Berkowitz (2018) in her seminal work on meta-organizations. Berkowitz (2018) found that a meta-organizing governance device can benefit a sustainable innovation ecosystem in cities by acting as a platform, has a cost-benefit structure and low transaction cost, consists of formal and informal organizations and stakeholders, results in self-regulation and system reconfiguration, decision making capabilities on a consensus, functions by heterarchy, inclusive of heterogenous actors and institutions, produces mechanisms for evaluation and reporting, facilitates knowledge transfers and collective adaptive learning, permits coopetition, competing institutional logics, and drives conversations, and produces information for member adoption and outreach. City ecosystems comprise of different associations of organizing inclusive of various stakeholders. In some parts of cities, we find that there is a facilitation of knowledge transfers and modulated networks of collective learning and decision making; however, it is rarely on consensus and usually modulated in centers of power and wealth within the ecosystem. Berkowitz also analyzed the organizational capabilities for fostering sustainable innovation and featured these meta-organization attributes that fostered one or more of the organizational capabilities (Berkowitz, 2018). Organizational capabilities are relevant to the organizational efficacy of fostering sustainable development in ecosystems; however, I posit that a macromanagement approach and dynamic capabilities go beyond the operationalized routines within ecosystems by creating new routines and resource configurations for ecosystem sustainable development (Pavlou & El Sawy, 2011; Teece & Leih, 2016). I acknowledge the six organizational capabilities because I find them useful to at least identify and incorporate them into my conversation. The six organizational capabilities included *anticipation* of regulatory or institutional change, planning through coopetition and platform dimensions; *reflexivity*, examining limitations and trade-offs within the system through workshops, information production, and outreach activities; *inclusion*, including a wide range of actors and multiple stakeholder, as well as geographic inclusion; *responsiveness,* being flexible and offering tools to facilitate whole systems change; *accountability*, producing norms of ownership, self-regulation, reporting, and oversight mechanisms; and *resilience*, collective learning, knowledge transfers, and information pooling for adapting to perturbation by fostering innovative capacities (Berkowitz, 2018: 423).

In conclusion, it is the combination of collective action among multiple and diverse stakeholders and a meta-organizing governance structure that can help develop meta-organizing capabilities better than individuals and individual organizations do on their own. In the following sections, I will provide some insight into how developing dynamic capabilities and attributes of a meta-organizational governance device fosters sustainable development in city ecosystems. I will illustrate and provide a descriptive and prescriptive of how city ecosystems can develop towards flourishing CASE by the macromanagement approach, a meta-governance device, and dynamic capabilities development.

**The Dynamic Capabilities Lens for Developing Ecosystem Innovation**

The dynamic capabilities (DC) lens is based on the resource-based view (RBV) developed by Teece et al. (1997) and Eisenhardt and Martin (2000) within the management and strategy literature streams. DCs have been defined as the organizational and strategic routines to achieve new resource configurations (Eisenhardt & Martin, 2000). I define DC within the context of the theoretical systems approach to management, where managers and decision makers in the system develop new capabilities to better match its environment for competitive advantages (Teece, 2018b). The DC lens can focus on how systems can adapt to the changing socio-economic and socio-ecological environments by reconfiguring resources and capabilities (Eisenhardt & Martin, 2000). Although some scholars have criticized the DC lens for being tautological reasoning or non-operational (Priem & Butler, 2001; Williamson, 1999), I approach DC to highlight their capacities to identify opportunities for systematic reconfiguration, to develop a strategic response to certain needs, goals, or opportunities, and to implement courses of action (Helfat, Finkelstein, Mitchell, Peteraf, Singh, Teece et al., 2009). These capacities determine what the organization, or meta-organization in my case, can do and how effectively it makes changes.

Teece (2007) noted, "dynamic capabilities can be disaggregated into the capacity to 1) sense and shape opportunities and threats, 2) to seize opportunities, and 3) to maintain competitiveness through enhancing, combining, protecting, and, when necessary, reconfiguring the [systems'] intangible and tangible assets" (p. 1319). The sensing, seizing, and reconfiguration capabilities are the underpinning processes and micro-foundations of DC that extend beyond the operational routines and capabilities of systems (Pavlou & El Sawy, 2011). For example, these underpinning processes can include R&D, technology and/or knowledge transfer routines, alliance and acquisition capabilities, and resource allocation routines (Eisenhardt & Martin, 2000; Teece, 2007).

Sensing capabilities is essentially the adaptive learning process of identifying opportunities (Teece, 2007) while seizing capabilities are the orchestrating processes of identified opportunities to develop and deliver innovative value propositions (Teece, 2020). The reconfiguring capabilities are aligned with the competitive nature of the system over time by adapting resources and structures to changing external environments (Teece, 2007). As Teece (2018b) argues, "[As a systems management approach] the goal is not short-term efficiency, as in classic management, but rather the maintenance of 'evolutionary fitness' over time" (p. 363). I apply the DCs in the framework of innovation ecosystems within city ecosystems; because in the context of city-level sustainable development, it provides a lens that enables innovation ecosystems to develop this 'evolutionary fitness' of city ecosystems over time. While Teece (2007) developed this concept of 'evolutionary fitness,' other scholars have described this effect as adaptive capacities towards building resiliency in organizational systems and communities (Bec et al., 2019; Dale et al., 2010; Gulati, 2010). Since my overarching focus is on the city ecosystem, the framework extends the lens from just the firm and firm resources, but expands the framing to include external firms, institutions, and stakeholders within the city ecosystem processes.

Before I continue, I must define innovation ecosystems and explain their relationships and relevance to city ecosystems in the context of sustainable development. According to the literature, an innovation ecosystem can be defined as the "alignment structure of the multilateral set of partners that need to interact in order for focal value propositions and shared goals to materialize" (Adner, 2017: 42). As it relates to developing cities as sustainable ecosystems and smart cities, in particular, innovation ecosystems link and leverage information technologies (e.g., sensors and connected devices, open data analytics, platforms, and fiber-optic networks) with human capital (e.g., universities, companies, and public institutions) to foster sustainable innovation for a shared vision of achieving sustainable development goals (Angelidou, 2014; Appio, Lima, & Paroutis, 2019; Berkowitz et al., 2020; Sachs et al., 2019). Underpinning this linkage and leverage in smart sustainable cities are the processes of improving urban performance and form by utilizing socio-technical devices to provide more efficient services to citizens, to monitor, optimize, and upgrade existing infrastructure, to increase collaboration amongst different economic actors and to encourage innovative capacities in both private and public sectors (Marsal-Llacuna, Colomer-Llinàs, & Meléndez-Frigola, 2015). This process requires a high level of resource orchestration and leadership.

One of the distinguishing features of an ecosystem is the presence of a "powerful actor" or set of "powerful actors" for resource orchestration. These actors, also considered as some form of stakeholder leadership, set the system-level goals and establish members' roles, standards, and interfaces (Adner, 2017; Gulati et al., 2012; Schoemaker et al., 2018). This requires a collection of skills in forging and maintaining partnerships (Ginsberg, Horwitch, Mahapatra, & Singh, 2010; Li & Garnsey, 2014), managing technology infrastructure (Adner & Kapoor, 2010; Almirall, Lee, & Majchrzak, 2014; Gawer & Cusumano, 2014), governing the ecosystem (Adner, 2017; Berkowitz, 2018), and managing value-creation and value-capture activities (Kapoor & Lee, 2013). Since this orchestration requires a complex strategic approach and action, some scholars have acknowledged resource orchestration as a dynamic activity and have recognized it as orchestration capabilities (Hurmelinna-Laukkanen & Nätti, 2018; Linde, Sjödin, Parida, & Wincent, 2021; Teece, 2020). As Verhoeven and Maritz (2012) describe, "…innovation ecosystem orchestration is the set of deliberate, purposeful actions undertaken by organizations for initiating and managing innovation processes in order to exploit marketplace opportunities" (p. 5). I challenge the notions from previous studies and position my argument aligned with Berkowitz (2018) and Cooperrrider (2012). As they argued, the innovation ecosystem orchestration should not depend on a single powerful leader with hierarchal governance structures and management approach but needs a heterarchical governance structure in core leadership consisting of several powerful leaders with a diversity of strengths, resource power, and clout to foster innovation within the ecosystem.

Within this section, I narrowed a descriptive conceptual lens of DC from the city ecosystem down to the orchestration capabilities of the core leadership. In doing so, I have acknowledged that DCs exist within city ecosystems, as they integrate the resource orchestration that enables innovation ecosystems to continuously adapt and maintain a CASE as well as form the ecosystem's evolutionary fitness. The literature is well established on the role of DC development in firms. However, emerging streams of scholarship are exploring the role of DC in structuring and managing ecosystem innovation. Some scholars have explored specific roles of DC for service innovation within the energy utility sector (Lütjen, Schultz, Tietze, & Urmetzer, 2019), while other scholars have studied how DC can guide universities through resource orchestration within local innovation ecosystems (Heaton et al., 2019). Feng, Fu, Wei, Peng, Zhang, and Zhang (2019) looked at the role of DC in helping start-ups develop into ecosystem leaders by designing an evolutionary framework for the start-up process. The potential of dynamic capabilities to increase value creation and capture for digital platforms leaders was studied by Helfat and Raubitschek (2018). They developed several critical propositions for leadership in ecosystem orchestration to consider for value creation, which included innovation capabilities, environmental scanning and sensing capabilities, and integrative capabilities.

I collectively combined all of these capabilities as the underlying processes and routines that enable DC within the ecosystem-innovation context. More specifically, I highlighted the resource orchestration DC, the sensing DC, the seizing DC, reconfiguration DCs. These DCs are arguably the requirements for city ecosystems to remain competitive over time by its stakeholders' finding ways of applying the diversity of capabilities together in creating sustainable value within ecosystems. This includes private and public enterprises, service innovations, social, cultural, and product innovations that benefit the city ecosystem by creating a flourishing environment.

As I imply the benefits of "evolutionary fitness" and competitive nature in city ecosystems, I do not suggest city ecosystems compete as firms do by profit seeking, pricing, evolving, market share, surviving, and dying (Eisenhardt & Martin, 2000). I do, however, argue that certain cities "compete" by having competitive advantages over others as it relates to the socially desirable amenities (Florida, 2002). Such competitive advantages include the geographic location, social, and cultural capitals, an attractive built environment with accessible transportation options, access to good education and paying jobs, and thriving health and business services of a city ecosystem (Flora, Emery, Fey, & Bregendahl, 2005; Florida, Mellander, & Stolarick, 2008). As some scholars have suggested, it is the constant orchestration of the current city ecosystem's built environments and human capital innovative capacity along with a city ecosystem's attractiveness to draw new talent and skill sets to incorporate into the local ecosystem (Florida, 2014; Katz & Wagner, 2014).

Very few studies have investigated the interdependent relationships of ecosystem related DCs and their development in fostering sustainable innovation for achieving CASE. By understanding ecosystem dynamic capabilities with the collective processes of innovation, and how the city ecosystem's leadership approaches resource orchestration within the ecosystem can be integrated to reveal a more comprehensive appreciation of how ecosystem leaders and decision makers can best develop dynamic capabilities to foster sustainable innovation that benefits the city as a whole ecosystem. From my literature search, I have identified a set of ecosystem-level dynamic capabilities that are linked to fostering ecosystem sustainable innovation in cities. Specifically, I focus on the work by Linde et al. (2021) that uncovered three ecosystem DCs which enable ecosystem orchestration capabilities to foster innovation. They then concluded that a combination of sensing, seizing, and reconfiguring enables three innovation orchestration mechanisms that develop the orchestration capabilities for fostering innovation in city ecosystems. Table 3 provides a descriptive of these ecosystem capabilities as it relates to orchestrating ecosystem innovation that benefits cities' ecosystems.

**Table 3: Literature Synthesis of Ecosystem Innovation and Dynamic Capabilities**

| Resource Orchestrating Capabilities of Ecosystem Leadership | Ecosystem Dynamic Capabilities | Description |
|---|---|---|
| <u>Orchestrating Capabilities</u><br><br>1. Configuring ecosystem partnerships ensures that firm leaders can direct their ecosystems to achieving evolutionary fitness through environmental changes | Sensing | Enables stakeholders in ecosystems to explore novel business, service, and cultural opportunities; and forge potential ecosystem partners for collaborating and integrating resources. Expanding physical and geographic limits by digitalization advantages |
| 2. Deploying value propositions ensures the development of new value propositions with multiple ecosystem actors, organizations, and sectors<br><br>3. Governing ecosystem alignment exercises the critical function of finding sustainable partnership configurations where complementary assets are shared and leveraged across firms, and benefits and costs are fairly distributed | Seizing | Stimulates the ability to realize and exploit (cultivate) opportunities through developing public and private commercially attractive value propositions that are also conscious of ecological and social indicators of sustainability-as-flourishing |
| | Reconfiguring | Ensure that the ecosystem is adaptive and flexible to changing external and internal conditions and with the ability to reorganize its relationship structures and digital offering to better suit changed conditions in the interests of long-term competitiveness and ecosystem evolutionary fitness |

Previously in this literature review, I argued ecosystem resource orchestration should be initiated from a set of powerful actors or leadership and not a single powerful actor within a city ecosystem. In doing so, the diversity of leading orchestrators starts the preconfiguring of ecosystem partnerships, ensuring that firm leaders can direct their innovation ecosystems to achieving evolutionary fitness through social, economic, and ecological environmental transformation (Linde et al., 2021). By the deployment of the grand value proposition, or shared goal(s), the heterarchical leadership ensures and disseminates the shared sustainable developmental goals so that the co-creation of new value propositions can be explored with multiple city ecosystem actors, organizations, and sectors (Linde et al., 2021; Sachs et al., 2019). Lastly, the core leaderships' governance structure of ecosystem alignment extends the function of forging sustainable partnership configurations. This is where complementary resources, strengths, and capabilities are shared and leveraged across partnering organizations and citizens, and benefits and costs are fairly distributed (Berkowitz et al., 2017; Cooperrider & McQuaid, 2012; Laszlo & Cooperrider, 2010; Linde et al., 2021). Since the complementary resources, strengths, and capabilities are shared, and decision-making contributions are considered interorganizational, resource orchestration capabilities are further developed ecosystem wide.

Furthermore, I argue that the underlying processes of developing ecosystem dynamic capabilities for innovation are initiated by ecosystem leadership orchestration, but the city ecosystem's dynamic capabilities can be fully developed through meta-organization. For example, ecosystem sensing capabilities enable the diversity of stakeholders in city ecosystems to explore and learn about novel business, service, and cultural opportunities with sharing of indigenous and technical knowledge as well as the sharing of industry knowledge of competing private organizations and competing institutional logics (Almirall et al., 2014; Teece, 2018b). As participating stakeholders and organizations forge more partners within the ecosystem, they create dialogue opportunities for the planning, collaborating, and integrating resources. Because there are internet, broadband, and other communication advantages available in cities, this allows the multitude of stakeholders to sometimes span boundaries of physical and geographic limits and collaborate with higher frequencies (Adner, 2017; Almirall et al., 2014).

The ecosystem seizing capabilities stimulates the ability to realize and exploit (cultivate) opportunities through developing public and private commercially attractive value propositions that are also conscious of ecological and social indicators of sustainability-as-flourishing

Ecosystem reconfiguring capabilities ensure the ecosystem's adaptivity and reflexivity, or resiliency, by the co-created changing of external and internal conditions. Ecosystem reconfiguring capabilities also foster the ability to reorganize its relationship structures and digital offerings to better suit the changed conditions (Linde et al., 2021). For example, in the energy sector's innovation ecosystem, organizations, regulators, value adding, and non-value adding partners can share and realign knowledge-transfers to adapt cross-industry and to consumers (Lütjen et al., 2019). In addition, by establishing a meta-organizing governance structure for the city ecosystem innovation, relational structures between (value-adding) producers, suppliers, and so on, and assumed non-value adding (consumers) partners, the ecosystem is maintained and enhanced by continuous knowledge transfers in the interests of long-term ecosystem evolutionary fitness (Lütjen et al., 2019; Rindova & Kotha, 2001; Teece & Leih, 2016; Tsou & Chen, 2020).

Hence, a meta-organizational governance structure for innovation ecosystem development of dynamic capabilities fosters innovation within the city ecosystem, principally in the case of sustainable innovation. To better understand why developing dynamic ecosystem capabilities by a meta-governance structure, I relate the attributes of meta-organization with ecosystem sustainable value creation from ecosystem dynamic capabilities development. Table 3 presents a synthesized view of how meta-organizational attributes can develop ecosystem dynamic capabilities for innovation ecosystems in the context of CASE.

I have identified the attributes of meta-organizations as a governance device for city-level ecosystem sustainable development (Table 2). I have also uncovered four dynamic capabilities needed for developing ecosystem innovation towards flourishing CASE (Table 3). Table 4 presents the synthesized view of how meta-organization attributes can develop dynamic capabilities; thus, validating my case that meta-organization is a useful governance device for fostering dynamic capabilities that produce sustainable innovation in ecosystems. The synthesized summary of how the meta-governance device attributes foster ecosystem dynamic capabilities is explained as follows.

## Table 4: Relating Ecosystem Sustainable Innovation to Meta-Organization Attributes

| Ecosystem Dynamic Capabilities | Meta-organizing attributes | Citation |
|---|---|---|
| Orchestration – *Formation of ecosystem leadership and ecosystem resources* | – Creating workshops and platforms for collaboration<br>– Coordinating and integrating partnerships | (Adner, 2017; Berkowitz, 2018; Linde et al., 2021) |
| | – Producing norms of ownership, self-regulation, reporting, and oversight mechanisms | (Berkowitz, 2018; Linde et al., 2021; Lütjen et al., 2019) |
| | – Including a wide range of resource-rich actors and diversity of multiple stakeholders, sectors, formal and informal inclusion, and low-cost structure | (Adner, 2017; Berkowitz, 2018; Ciborra, 1996; Gawer & Cusumano, 2014; Linde et al., 2021) |
| Sensing – *Partnership scouting and outreach, identify strengths and opportunities* | – Heterarchical planning, strategizing, and co-creation through coopetition, competing institutional logics, and platform dimensions.<br>– Continuous sourcing and forging of potential partnerships | (Berkowitz, 2018; Cooperrider & McQuaid, 2012; Linde et al., 2021; Lütjen et al., 2019; Teece, 2018b) |
| | – Identify and examine opportunities, collective adaptive learning through multistakeholder conversations, knowledge pooling | (Berkowitz, 2018; Cooperrider & McQuaid, 2012; Linde et al., 2021; Lütjen et al., 2019) |
| | – Collective learning, knowledge transfers, and information pooling for adapting to perturbation by innovative capacities | (Berkowitz, 2018; Bogers, Chesbrough, Heaton, & Teece, 2019; Cooperrider & McQuaid, 2012; Linde et al., 2021; Lütjen et al., 2019) |
| Seizing – *Value proposition strategic development; ecosystem formation, coordination, and integration* | – Cultivating (Exploiting) triple bottom line opportunities, adapting to limitations and trade-offs within the ecosystem through information production, and outreach activities | (Berkowitz, 2018; Linde et al., 2021; Lütjen et al., 2019) |
| | – Consensus-based decision making<br>– Co-development of sustainable value creation strategy<br>– Formation of strategic plan and designing value propositions | (Adner, 2017; Berkowitz, 2018; Kapoor & Lee, 2013; Linde et al., 2021; Teece, 2018b) |
| | – Balancing collaborators and competitors to distinguish private and public business modeling and development | (Lütjen et al., 2019) |
| | – Controlling ecosystem bottlenecking and intermediating entangled logics | (Lütjen et al., 2019) |
| Reconfiguration – *Adaptive value co-creating and implementing; ecosystem resilience* | – Anticipatory value proposition co-creation by prospection and wholeness mindset | (Berkowitz, 2018; Kapoor & Lee, 2013; Laszlo & Cooperrider, 2010; Linde et al., 2021; Teece, 2018a) |
| | – Being flexible and offering tools to facilitate whole systems change<br>– Building resilience in the ecosystem | (Berkowitz, 2018; Laszlo & Cooperrider, 2010; Linde et al., 2021; Lütjen et al., 2019) |
| | – Regulatory or institutional change making<br>– Adaptive and flexible to changing internal (organizations and stakeholders) and external (ecosystem) conditions for evolutionary fitness | (Berkowitz, 2018; Linde et al., 2021; Lütjen et al., 2019; Teece, 2018b) |

*Orchestration.* Orchestration dynamic capabilities are defined as the formation of ecosystem leadership and ecosystem resources. Rare instances of leadership types can include a single leader with managerial dynamic capabilities (Adner & Helfat, 2003). Because this single actor would have a respectable level of power, clout, and connected to a bevy of resources, Teece (2018b) would consider this leader to be the valuable, rare, imperfectly imitable, and non-substitutable (VRIN) element within the ecosystem. But many city ecosystems consist of varying leadership models and require the orchestration of leadership power, resources, and capacities to meet the higher-level ecosystem dynamic capabilities development. Therefore, orchestration DCs require the decentralization of authority, and the heterarchy of sectoral leaders in business, government, and community create partnerships amongst themselves as they become the ecosystem's nucleus, or core ecosystem leadership. Then the core ecosystem leadership enables further partnering internally throughout and external to the ecosystem (Adner, 2017; Berkowitz & Bor, 2018). There are some examples of core ecosystem leadership initiating, coordinating, and integrating partnerships in the development of orchestration DC, particularly in developing sustainable city ecosystems. Heaton et al. (2019) studied the central role of universities functioning as a catalyst for sustainable city development by creating platforms and meta-organizing programs for collaboration in innovation ecosystems. MacDonald, Clarke, Ordonez-Ponce, Chai, and Andreasen (2020) and Berkowitz (2018) observed the resource orchestration of municipal sustainability managers playing a central role in cities and their efforts for creating programs inclusive of producing sustainability norms, fostering sustainable innovation, and creating partnerships for regulatory, reporting, and oversight mechanisms for sustainable development. The higher-level activities and routines of this meta-organization are responsible for the design of collaborative workshops and acting as platforms. One of the key parameters of meta-organization and resource orchestration for sustainable innovation is to involve a wide range of resource-rich actors and diversity of multiple stakeholders, sectors, as well as formal and informal inclusion; and to plan and create a decentralized, low-cost structure by pooling the resources available within the ecosystem partnerships (Berkowitz, 2018; Linde et al., 2021; Lütjen et al., 2019).

*Sensing.* Sensing dynamic capabilities are the first of two parts of the adaptive learning process. It is defined as the continuous partnership scouting and outreach for expansion potential partnerships and efforts. This occurs by creating an inclusive environment with the meta-organization governance structure. Current members, as well as newly formed organizational partners, work together in platforms, innovation programs, and workshops to identify resource strengths and opportunities. The participating members of the innovation ecosystem learn from both technical and informal indigenous knowledge through conversation and knowledge exchange (Berkowitz et al., 2017; Bosch-Sijtsema & Bosch, 2015; Lütjen et al., 2019). A feature of sensing DCs is that the collective learning, knowledge transfers, and information pooling results in the primary phase of the adaptive learning process where members gain new knowledge from the diversity of sectors. This phase is critical to planning for sustainable innovation as responses to perturbations and design problems in the ecosystem. By forging relationships with competing institutional logics, fusing formal and informal knowledge together, and promoting coopetition within industries allow for different perspectives and problem solving approaches to occur; which may also lead to the emergence of unique solutions or capabilities that may not have been considered independently from fewer or a single perspective (Berkowitz & Dumez, 2016; Li & Garnsey, 2014; Linde et al., 2021; Lütjen et al., 2019).

*Seizing.* Seizing dynamic capabilities is the second phase of the adaptive learning process, where value proposition strategic development, ecosystem coordination and integration occurs. New partnerships continue to form as outreach continues, and the new partnering organizations are integrated into the adaptive learning platform. By exploiting (cultivating) triple bottom line opportunities learned from conversations, members find ways of adapting to limitations by considering the qualitative and quantitative trade-offs within the ecosystem through information production and outreach activities (Berkowitz, 2018; Lütjen et al., 2019). Features of seizing dynamic capabilities include consensus-based decision-making, the formation of the strategic plan, and designing value propositions for sustainable innovation. However, there are challenges in the integration process of seizing. For example, members must work through controlling ecosystem bottlenecking and intermediating entangled logics when they occur, especially when there are multiple institutional logics integrating and no intermediary institutional logic can be formed (Hamann & April, 2013; Lütjen et al., 2019). In addition, as the strategic plan is developed, collaborating members, which sometimes are inclusive of competitors, have to balance organizational and inter-organizational information sharing. Moreover, as new strategies, capabilities, and innovative artifacts are realized, there also has to be a prospective sorting process to nest and distinguish private business modeling and development from public offerings before the action and implementation stages of development.

***Reconfiguration.*** Reconfiguration dynamic capabilities consist of adaptive value

co-creation and implementation within the ecosystem. Reconfiguration DCs increase

innovative capacities for business purposes, and they contribute to the continuous

building of ecosystem resilience. Members in the meta-organization co-design value

propositions, prototypes, and simulations to test and question the impacts of emerging

technologies on social, economic, and ecological environments (Berkowitz, 2018; Linde

et al., 2021). From these emerging technologies, members can anticipate how these new

innovations can shape the evolution of the city ecosystem and what benefits to society the

innovation can yield in wholeness. This will include members anticipating future

regulatory and institutional changes to better adapt innovations, regulatory risks, and

policies to the ecosystem (Berkowitz, 2018). Being flexible and offering tools to facilitate

whole systems change is another important feature in reconfiguration DCs. For example,

for-profit organizations in the meta-organization can adapt their private value creations

by interacting with participating stakeholders. Together, they can co-develop sustainable

customer-based innovations from feedback learning loops, and in some cases, they can

adjust operational processes from the changing environments that benefit the triple

bottom line in that business and spills over into the city ecosystem (Adner & Kapoor,

2010; Hurmelinna-Laukkanen & Nätti, 2018; Linde et al., 2021). Moreover, start-up

platforms nested within innovation ecosystems have been found to play a key role in co-

developing value propositions and future ecosystem leadership (Feng et al., 2019). As

Feng et al. (2019) suggest, "This development is in prospection of potential changes in

social and ecological environment challenges city ecosystems may face, but also

preconfigures start-ups and their innovations as the future leaders in continuity of holistic evolutionary fitness within the ecosystem" (p. 94). In conclusion, reconfiguration DCs

are the reorganizing, adapting, and responding to the changing internal (organizations and

stakeholders) and external (ecosystem) conditions. The continuous organizing and re-

organizing of specific tangible and intangible assets fosters new value propositions in the

form of sustainable innovation that develops business opportunities and growth while

also extending benefits to societies in cities as a whole. Reconfiguration of the whole

systems framework embraces the integrated management of complementarities; and

increases adaptive capacities with the utilization of co-specialized assets for evolutionary

ecosystem fitness (Adner & Kapoor, 2010; Teece, 2007).

**Meta-theorizing Integrative Management for City Ecosystem Flourishing**

Tying it all together, a metatheoretical approach for ecosystem sustainable

development that integrates macromanagement, dynamic capabilities, and meta-

organization lenses has the potential to resolve fundamental and deeply rooted paradoxes

in the study of sustainable development towards ecosystem flourishing. In this final

section of the literature review, I will explain how the macromanagement, meta-

organization, and dynamic capabilities lenses are used to consider the city ecosystem

sustainable development paradox between an ecosystem and its interchanging parts, or

more specifically, the holism-reductivism paradox. The holism-reductivism debate is in

most scientific disciplines. As it relates to city ecosystems, systems dynamics has been

suggested as the key to resolving the tensions between holism and reductivism (Becht,

1974; Bergandi & Blandin, 1998). System dynamics combines theory, methods, and

philosophy to analyze systems' behavior (Forrester, 1998). The systems dynamics lens in

cities was further developed in J. Forrester's "Urban Dynamics," and in this work he proposed that physical and social systems are interconnected. Because the relationships

of the two should not be (Forrester, 1970). He also articulated,

> We do not live in a unidirectional world in which a problem leads to an
> action that leads to a solution. Instead, we live in an on-going circular
> environment. (Forrester, 1998: 2)

Here, I abduct from Forrester two critical conclusions that I anticipate my studies will

validate in my book: (1) The holistic lens is important for understanding management

approaches for highly complex ecosystems; and (2) Linear problem solving has not been

effective in city ecosystem development and community change management; therefore,

the complex challenges in city ecosystem development require complex problem-solving

such as an integrative ecosystem management approach.

"Urban Dynamics" (Forrester, 1970) was central to Boston's urban development

planning and design since the early seventies, and since then, has spawned several

research projects and city development projects that have influenced urban policy

decisions. Moreover, this book was instrumental in motivating the research in city

development of the book, as cities are now evolving from survival and efficiency

developmental goals to sustainable and flourishing development goals. To reduce any

confusion, I want to separate my integrative meta-theory of ecosystem development from

the functioning of my integrative ecosystem management approach.

Integrative systems management (ISM) in sustainability can be broadly defined as

a holistic approach with vertical and horizontal alignments of sustainability-related

activities that combine reductive management approaches with non-trivial human

behavior through inclusiveness, adaptiveness, and evolutionary transformation (Sroufe,

2018; Walker, 2013). The ISM perspective acknowledges wholeness instead of compartmentalized strategies and activities by aligning complementaries of capabilities

for ecosystem transformation. Macromanagement strategies vertically and horizontally

align the ecosystem's human resources, or stakeholders, for positive future-focused

collective action towards sustainability (Cooperrrider, 2012; Fredrickson & Losada,

2005; McFarland, 1977; Seligman et al., 2013). Meta-organization as a governance

device structures organizing organizations into a platform for knowledge pooling,

decision making, measuring, evaluating, and holding collective members accountable for

creating sustainable ecosystem innovation and giving birth to new value propositions in

the ecosystem by dynamic capabilities (Berkowitz, 2018; Linde et al., 2021; Visnjic,

Neely, Cennamo, & Visnjic, 2016). Dynamic capabilities determine what those involved

in ecosystem management are able to do and how effectively they can make change

(Teece, 2018b). Combining these functional elements in an ecosystem creates this

'continuous morphing' and illuminates the paradoxes between innovation and resilience,

growth and maintenance, which advances the evolutionary fitness of a city ecosystem

(Holling, 2001; Rindova & Kotha, 2001; Teece, 2018b). Therefore, the integration of

macromanagement, meta-organization, and dynamic capabilities are internal factors,

activities, and components that can influence a developmental identity of sustainable or

flourishing in individual human systems as well as the developmental identity of the

external whole of the ecosystem (Edwards, 2005, 2009; Sachs et al., 2019). Thus, an

integrative systems-based approach towards developing CASE cultivates ecosystem

resources, for example, human, political, built, social, natural, financial, and cultural, so

that the ecosystem transforms into higher levels of sustainable identity (Edwards, 2009;

Flora et al., 2005). Sroufe (2017) discovered similar findings in his work on integration

and organizational change towards sustainability. Sroufe (2017) suggests:

> …positive capacities towards the development of sustainable operating and enterprise-systems integration are required to develop integrated organization… [It is the combination of driver, enablers, and evaluators that positively impact change management [by] novel improvements, design, innovation, and stakeholder engagement… Integrated organizations will outperform rivals with less integration… (pp. 324–325)

There are strengths in employing an ISM for ecosystem sustainable development

because it draws upon different yet complementary concepts and theoretical frameworks.

However, transdisciplinary approaches have often been criticized for being too broad,

which runs against the grain of modern academia, particularly in management and B-

schools (Teece, 2011).

In academia, managerial, and professional practices, technical experts are

rewarded for deep specialization in merit-based ecosystems, which shape the way these

experts think. But this leads to partial system views and incremental change management

approaches. This classical management view influences human bias in actions and

systems of control, which are translated from their adaptive learning capacities, reductive

reasoning, and knowledge acumen (Doniger, 1999; Ruse, 2005). Classical management

entrenched in reductivism can cause a partitioners "blind spot" in the ecosystem by acting

separately from the ecosystem, problem solving from compartmentalized and flawed

information. For example, as a response to a blighted neighborhood, a developer won a

government bid to construct a building consisting of spaces for small business offices,

commercial spaces, and community grocery. During the planning, strategy, and

implementation stages of this community development project, there was no evidence of

dialogue within the community to address the needs of such a project, and there were minimal strategic alignments and efforts to bring businesses to occupy the building.

Neither were there any programs in place to provide training and other resources for

community stakeholders to develop small businesses to utilize the building. Needless to

say, this building remained empty, and the neighborhood is still blighted. However, the

project contributed to the short-term survival of the development company -at least by

economic means. This example shows how separate actors and organizations within the

system think by separateness and reductive reasoning and act separately from the

ecosystem with reductive approaches instead of a part of the ecosystem. The developers'

separated, limited, and linear way of thinking indicates that if he builds a structure—

which building is his specialization—for his and others' economic prosperity, then people

will autonomously come to fill the building with their businesses. Thus, by reductive

views and short-sighted "improvement" projects in the blighted community, the

developers missed opportunities for a system-focused strategic shift with far more

important long-term implications to the community and city ecosystem.

I shared this example for several reasons. First, to highlight separateness and

failures in reductive approaches to ecosystem development. The developers, and probably

the municipal planning agency as well, missed the human elements and socio-

psychological characteristics of the community project while only focusing on their

technical expertise in addition to linear reductive assumptions and conclusions. Second,

to stress the importance of integrating various ecosystem components in a platform, such

as businesses, social networks, institutions, and governments, to operationalize

sustainable development in city ecosystems. Third, to further unwrap a deeper sustainable

development for growth paradox nested as aggregated reductive linear forces and transformational nonlinear power in relevance as it relates to city ecosystem change

management approaches to flourishing.

In many Western cities, classical management heavily exists in reductionism and

has been the dominant force for understanding and implementing strategic developmental

approaches. Reductionism, as it relates to change management approaches to

development, is considered a linearly deterministic adaptive processing that reduces

complex behaviors to a simple set of variables that offer probabilities of solutions for

systematic change (Doniger, 1999). To be clear, I am not suggesting replacing

reductionist management with integrationist management because focal knowledge and

specialization have value in problem-solving. Additionally, the translational classical

change management approaches are deeply held within the institutionalized value

systems of modern societies (Edwards, 2009). However, solutions to complex sustainable

development challenges will probably not be achieved on the scale of a city ecosystem

with incremental and aggregated linear adaptive capacities of change management.

Moreover, the translational nature and adaptive capacities of reductive progression will

probably yield, at best, the developmental stage identities of compliant or efficient (refer

to Figure 2). As Jensen (2005) suggests, "Only generative transformational change

management, which require 'frame-breaking' insights and collective behaviors to be

experienced and institutionally implemented, can result in such [ecosystem]

transformations" (p. 63).

Jensen's notion is in concurrence with other scholars that have theorized

aggregated series of reductive forces that heavily focus on the management of economic

growth translates into prosperity (Edwards, 2009). For example, on the one hand we have the institutional forces of exploiting and extracting resources to generate wealth in our economic and political systems. On the other hand, there are competing institutional forces that pressure business and political systems to act more responsible within ecosystem production and consumption that also considers social and ecological environmental disorder (Edwards, 2009: 199). But this paradox between institutional forces has led to excessive translational growth, social dislocation and inequities for some, and isolated pockets of prosperity for others fragmented in city ecosystems. This is at the root of the sustainability paradox, where tensions collide as one institutional force pushes against the other, and in most cases, the stronger institutional force stabilizes and maintains the ecosystem's identity in the translational development forms of compliant and efficient. The stronger translational force preserves the hegemony of economic values over other types of values and defends material wealth creation over other forms of well-being. At best, sustainable development and the innovation that enforces this perspective and understanding of growth drives ecosystems (and the businesses within them) in horizontal growth trajectories translated into efficiency and compliance forms. At worst, ecosystems and nested business systems are organized and stuck in backward thinking and pre-conventional survival forms of resource orchestration operating with survival or avoidant identities in developmental economic growth translations. In many cases, these developmental identities limit adaptive capacities within the ecosystem, and in some cases, these forms of translational growth have been seen as a distraction to the task of transformational development (Edwards, 2009).

Aligning my argument with other sustainable development scholars, I consider a complex complementary mix of translational and transformative dynamics are instrumental in solving the ecosystem paradoxes of sustainable development and innovation with maintaining resiliency in the ecosystem. This would include a wholeness perspective, understanding, and change management approach of both, an internal-external property of the city ecosystem, and the horizontal-vertical alignments of human resources within social systems.

The integration of macromanagement, meta-organization, and dynamic capabilities shares a theoretic contextual commonality of wholeness and transformation. This novel combination may be a useful instrument for leaders and managers to expand their understanding of integrated systems management but may also be a useful tool for fostering sustainable innovation, ecosystem sustainable development, and the transformation towards flourishing city ecosystems. One of the features of this integrated systems change management is that it institutionalizes by operating from a 'wholeness' framework of "power" sources instead of "force" sources. The integrated whole systems framework promotes an internal conscious transformation that dissipates and transforms the external environmental field, or city ecosystem (Rosado, 2008). To understand what I mean by "force" and "power" and distinguish some fundamental differences, force has linear energy motions, usually creating polarity and counterforces, while power has no motion yet generates energy from within a morphogenic field. It is important to note that power should not be confused with force. Power has no motion, as force has a trajectory; instead, it is a state of being. Hawkins (2014) explains this best in his work, *Power vs. Force*,

> Force always moves against something, whereas power doesn't move against anything at all. Force is incomplete and therefore has to be fed energy constantly. Power is total and complete in itself and requires nothing

from outside ... Force always creates counterforce... its effect is to polarize
rather than unify ... Power on the other hand, is still. It's like a standing field
that doesn't move. Gravity itself, for instance, doesn't move against
anything. Its power moves all objects within its field, but the gravity field
itself does not move ... Power gives life and energy—force takes these away
... Force must always be justified, whereas power requires no justification.
Force is associated with the partial, power with the whole. (pp. 134–135)

In order for city ecosystems to transition into both internal and external
sustainable identities, a more collective human consciousness of wholeness must be
realized and internalized, then dissipated throughout the panarchic systems (e.g.-
organizations, meta-organizations, and institutions) and on throughout the ecosystem.
This can occur with the utility of a macromanagement process and a governance structure
for developing dynamic capabilities to foster sustainable innovation. Instead of
conventional paradigms of understanding a fragmented ecosystem dominated by
disconnection and distant entities with classical problem-solving approaches, the
combination of these three key functions promotes interconnectedness by forging
partnerships and collaborative efforts to create new process, capabilities, and innovative
artifacts. Figure 5 illustrates a comparative representation between conventional systems
change management—the classical reductive approach, and post-conventional systems
change management—the integrative approach.

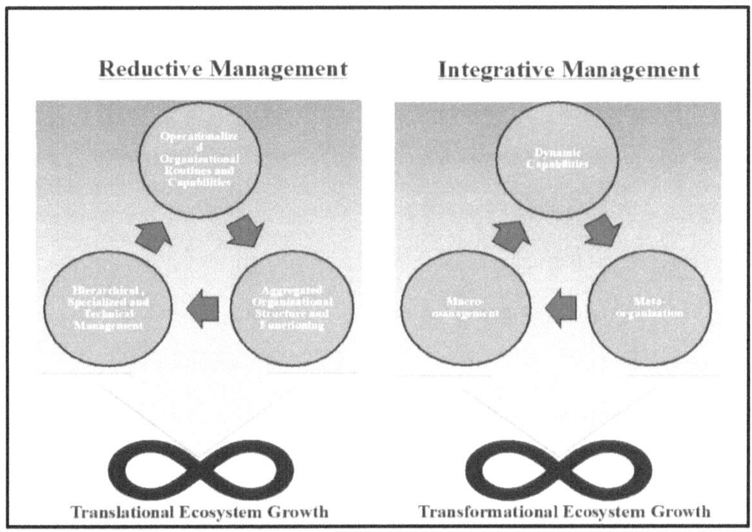



Classical change management with reductive approaches has functioned by top-down management forms of governance, decision-making, and strategic implementation. Highly specialized and technical experts have operationalized responsive routines specifically focused to their expertise and solve individual problems and threats that may occur in the ecosystem. These isolated social, economic, or ecological problems are approached with probable problem-solving solutions that may sometimes miss the full spectrum or causality of a particularly complex situation. Moreover, this management approach moves as forces of change on individuals and organizations, which many times results in counterforces such as push-back and revolt or other unintended negative externalities and consequences from social and ecological panarchic nested systems within the ecosystem (Holling, 2001). As the multitude of fragmented institutions, organizations, and managers continue to function as aggregated forces, the counterforces

continue to respond. This tension between forces develops poverty traps and rigidity traps

in the ecosystem, where "growth" metrics are not fully realized ecosystem-wide and

isolated at the top of the influence and affluence scales (Holling, 2001). Although this

change management approach is dominant in modern societies and city ecosystems,

aggregated reductionist approaches are limited to translational growth that may not reach.

One reason is that reductionist views of prosperity are heavily influenced and translated

mostly as economic value and fixed on the purpose of efficiency. The management

approach adapting the ecosystem heavily focused on economic value growth exacerbates

the city ecosystem traps by continuing to develop more organizational forces of

unchecked production and consumption mechanisms to drive economic growth over

other well-being metrics. In other words, stacking on a conscious ethos that translates

growth by economic prosperity only and doing "business as usual." This translational

growth limits a city ecosystem's developmental identity of "efficient" but flawed because

there is no true problem-solving. Problem-shifting comes into play, as business and

political systems justify "business as usual" by labeling social and ecological problems

that emerge as externalities while supporting the increase of productional routines for

profit. Even when city ecosystems identify as a developmental stage of "committed," I

find organizations, institutional properties, and cities with unsustainable practices and

bolted-on sustainable processes (Laszlo & Brown, 2014; Laszlo & Zhexembayeva,

2017). Even worse, "green-washing" and "blue-washing" in reporting mechanisms for

accountability has been identified in organizations and cities, which undermine

sustainable development by hiding accurate information instead of innovating improvements in businesses, models, and processes (de Jong, Han, & Lu, 2019; Delmas & Burbano, 2011; Macellari, Yuriev, Testa, & Boiral, 2021; Schuetze & Chelleri, 2016; Sur, 2017).

The advantage of post-conventional integration approaches for city ecosystems management is that they operationalize strategies that consider system wholeness and structure a meta-governance for heterarchical participation and decision making towards sustainable innovation for ecosystem transformation. Wholeness represents both internal and external factors of the ecosystem. Macromanagement, such as an Appreciative Inquiry platform that integrates top-down and bottom collective participation, reinforces positive generative outcomes in the city ecosystem by operationalizing a strengths-based whole systems process of planning and strategic implementation that is SOAR. Appreciative Inquiry also organizes organizations in a governance structure that pools knowledge, reductionist strengths, and promotes heterarchical decision making and designing opportunities for fostering sustainable innovation. Through meta-organization and macromanagement of the social systems, the dynamic capabilities (e.g., sensing, seizing, and reconfiguring) of the ecosystem are developed from resource orchestration to fostering new capabilities and innovations that create sustainability within the ecosystem. Combining these elements of the integrative ecosystem management process transforms both the human consciousness within the social system (internal) identified as sustainable or flourishing, and the designs of the city ecosystem mirror this consciousness of flourishing, achieving a transformational growth and ecosystem developmental identity (external). This transformation process represents the wholeness in systems flourishing. Figure 6 shows the variability of forms and identities that can occur over time as city ecosystems choose and balance transitional and transformational combinations of

management approaches.



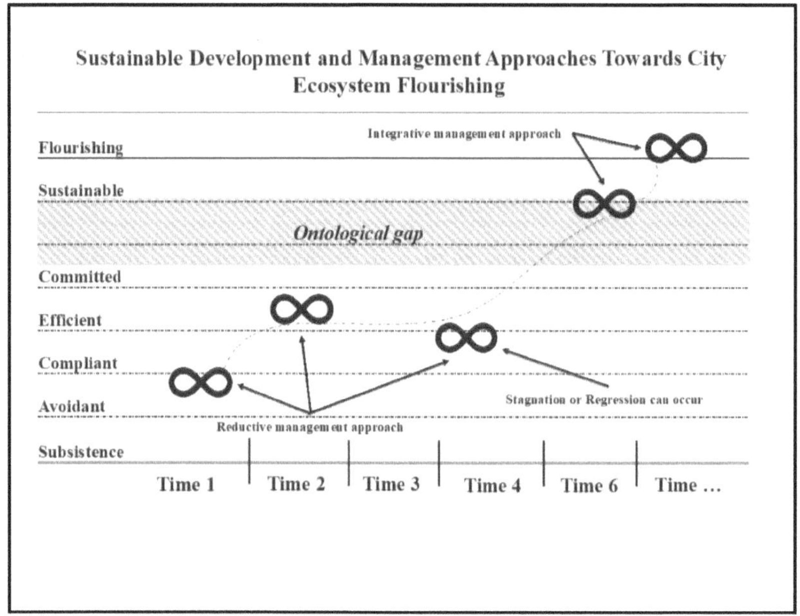


Reductive management approaches have linear translational growth models linked

to organizing and designing systems for the focal purpose of economic value.

Organizational systems and supporting institutions follow reductionist strategic

development that reflects a cities ecosystem's identity of efficiency with routinized

capabilities and little adaption to internal social and ecological environmental changes

over time. Moreover, organizational and institutional systems managers influence human

behavior within the social systems to sometimes act in a "money over everything" and

73

survival mode of thinking and acting that leads to innovative capacity building and entrepreneurship for making money efficiently, instead of other beneficial factors for human well-being. As this conscious mindset grows in ecosystems, the internal cohesions of the ecosystem deteriorate, and people become disconnected in their endless pursuit of economic value. What emerges are segmented organizing with various forces collaborating and competing for resources over time. Individual humans and collective humans in organizational systems can only move from developmental stages of survival to commitment to sustainability as they translate their prosperity on economic value and efficient organization. Since the compilation of human activities and interactions are reflections of the city ecosystem developmental identity, hence the overall identity of the city ecosystem. The accumulation of various reductive forces advances translational ecosystem growth, lacking organizational capabilities to resonate with transformative ecosystem growth. The ceiling to this development growth stage is "committed" because reductive adaptive capacities and approaches are forces of maintenance that do not have transformation power within the ecosystem (Edwards, 2009; Rosado, 2008). Stages of development can even regress to "compliant" or lower to "survival modes" as the gradient is linearly linked to time and money. External turbulences can emerge, which disrupts ecosystem development approaches and goals as adaptive responses were not considered or anticipated causing ecosystem regression. For example, a virus pandemic or an industry heavily embedded in a city ecosystem such as the automobile in Detroit may have dire consequences to socio-environmental concerns that temporarily disrupt the internal and external elements of the city ecosystem.

In order for city ecosystems to reach developmental stages of sustainability and flourishing, cities must utilize transformational power by adopting a consciousness of plurality and an integrated systems management approach. Integrative systems management helps ecosystem leaders, managers, and organizations cross what Pavez et al. (2020) called the "ontological gap of transformation towards the Path to Interconnected (PIC)" (p. 2). Only transformation power can achieve this paradigmatic mind-shift needed to holistically cross this gap and transition city ecosystems along with the businesses, organizations, institutions, and societies within them. Not only is this a conscious shift towards value systems of plurality and caring to create businesses and institutions for world benefit, but also to generate inclusion and integration in the much larger context encompassing human, social, cultural, political, and ecological systems' resources within city ecosystems (Flora, Flora, & Gasteyer, 2016; Pavez et al., 2020).

The connection between transformational power generation towards flourishing ecosystems and sustainability, in general, has been evaluated by many scholars (Hawkins, 2014; Sachs et al., 2019; Sasaki, 2014; Senge, 2003; Tsao & Laszlo, 2019; Winnard, Adcroft, Lee, & Skipp, 2014). To achieve the transformation towards sustainable organization—and meta-organization—as well as ecosystem sustainable development goals requires the adoption of new paradigms of consciousness and innovative behaviors. These changes probably cannot emerge without integrative systems change management for a whole systems conscious transformation. The lack of consciousness around valid transformational goals and the unchecked pursuit of translational compliance, efficiency, and productivity exacerbates the sustainable development conundrums (Edwards, 2009). As expressed by Dunphy (2003), no amount of translation results in transformation. It is the application of an integrative approach as a change management instrument that can address the paradox of sustainable development. Translational growth provides stability and adaptive capacities to sustain coherent, horizontally trapped identities. By accumulating more efficiency-based organizing principles and purposes of reductive forces only advance translational ecosystem growth, lacking organizational capabilities to resonate with transformative ecosystem growth. To achieve transformational growth in city ecosystems, the systems management approach must amplify the shift to more progressive and integrative states of sustainable and flourishing organizing and purposes (Edwards, 2009).

In this section, I highlighted the ecosystem sustainable development paradoxes by offering a conceptual solution of integrative systems management. In doing so, I unwrapped then combined the theoretical elements of macromanagement, meta-organization, and dynamic capabilities as complementaries that integrated my explanation to frame-breaking the reductive change management approach while propositioning the whole systems change management approach for flourishing city ecosystem transformation. The integrative systems management approach resonates with the development of strategic power to amplify a consciousness towards societal and city ecosystem flourishing. Finally, in this literature review, I provided the empirical evidence to support my unit theories and meta-theory as foundations to validate my research claims. I applied the micro-foundations within the unit theories of macromanagement, meta-organization, dynamic capabilities as internal descriptive and prescriptive elements that aided the application of the foundational structural functions (e.g., birth [$\alpha$], growth [r], maintenance [K], and opportunity [$\Omega$]) the adaptive cycle functioning (e.g., exploitation, conservation, reorientation release, and reorganization) of a CASE. The complex adaptive and responsive cycles of city ecosystems form the external developmental identities, which encompass the collective of human and organizational activities seeking growth, well-being, and prosperity within the ecosystem. City ecosystems' management that overwhelmingly embeds reductionist approaches for change produce translational growth identities in ecosystem development. City ecosystems' management that practices integrative systems approaches for change produce transformational growth identities for sustainable ecosystem development. I anticipate the generation of new knowledge and findings within the contexts of sustainable city ecosystems and ecosystem innovation by offering three empirical studies to approach my inquiry, then examine and validate my claims of the phenomena.

The meta-theoretical lens of ecosystem sustainable development framing will accomplish two things. This framing highlights a resolve between the holism-reductionism paradox, and it will support my claim that integrative ecosystem management approaches will be key to achieving sustainable city goals towards flourishing. The meta-theoretic focal lens was framed to explain how the internal forces and mechanisms within cities influence the external developmental identity of a city ecosystem. This framing allowed me to explain major patterns of moving parts within a city ecosystem in their developmental goals while also acknowledging the whole ecosystem's form. The unit theoretic lenses of macromanagement, meta-organization, and dynamic capabilities were used in framing the three studies of the book. As I illustrated in Table 1, the three studies provide different aspects of city ecosystem sustainable development. Each study was either uniquely framed in one or more of the unit theories or the theoretic lens emerged as a thematic construct in Study 1. I am

inspired by Tufte (1985), "The Visual Display of Quantitative Information," as he

advocated for merging multiple qualitative and quantitative information for better

analyzing, interpreting, and presenting data and research. For example, in Study 1, I

captured the lived experiences of sustainability managers and practitioners, which I

considered mostly as the "top-down" component of an ecosystem's implementing meta-

organization. From their lived experiences in sustainable development, I extracted meta-

organization and four other thematic constructs as critical factors for sustainable

development project success. In Study 2, I took the constructs from Study 1 and created a

model to conceptualize the utility of a macromanagement, meta-organizing platform that

enables the implementing meta-organization to develop dynamic capabilities for

innovation and resiliency in the ecosystem. Since Study 2 was an SEM approach, I was

able to capture a wider ecosystem stakeholder perspective that included a diverse

representation that participated in the implementing process of sustainable innovation and

development projects in cities. By capturing this broader sample of city ecosystem

stakeholders, I highlight the perspectives of both top-down and bottom-up stakeholders in

my empirical evidence. Study 3 was framed in the meta-organization and dynamic

capabilities lens. With the layering of the findings from my three studies, I first will

provide empirical evidence that acknowledges the importance of meta-organization

formation in integrative ecosystem management from Study 1. I anticipate that I will also

reveal the effectiveness of Appreciative Inquiry and the SOAR framework—which I

theorize as a macromanagement meta-organization platform—in the implementing

process towards building innovative capacities and resiliency for progressing and maintaining a flourishing ecosystems vitality. Study 3 was framed in meta-organization and dynamic capabilities lens, and the focal aspect of the city ecosystem was performance and what combination of factors contributed to the ecosystem performance. The empirical evidence in Study 3 will support the meta-organization lens and dynamic capabilities lens through the construct of powerful leadership asset orchestration. Powerful leadership asset orchestration has been suggested as a dynamic capability (Linde et al., 2021), and in my study, I acknowledge this factor as an important catalyst in the initial stages of ecosystem sustainable development implementation as the leadership within communities enables the linkages of organizations and systems, forms the shared goal, vision, or interdependent tasks, prioritizes the challenges to address unique to the city ecosystem, and the leadership also extends decision-making abilities amongst peripheral stakeholders. The arrangement of evidence in my three studies' results will answer each subsequent research question and primary research question that also supports the explanation of my meta-theory of ecosystem sustainable development. Moreover, I expect my three studies to support and advance my meta-theory that ecosystem sustainable and flourishing developmental identities are formed by a transformational integrative ecosystem management approach and are less likely to form by the sum of translational reductive management approaches.

In the following section, I will reiterate my primary research question and arrange the research questions from my three studies to discuss how the subsequent questions and empirical studies approach and support the primary inquiry. There are three research questions and research studies within my research design. I will explain my research design in detail by explaining the methodological approaches and attached theories from

my theoretical background framing and literature review. I will then present the three
research projects with an introduction, methodology, and discussions, and finally, present
an integrated conclusion to my mixed methods approach.

## CHAPTER 4: RESEARCH DESIGN

In this section, I reintroduce the primary research question. Thereafter, I introduce the supporting inquiries of Studies 1, 2, and 3. The theoretical framing lens provided within my literature review will also be presented along with a description of each study's research design. From the individual supporting studies, I propose that the findings and analyses of these three studies will provide empirical evidence that supports answers to each supporting inquiry. I expect the combination of the three studies should support an integrated explanation that solves my primary inquiry and generates knowledge for the meta-theory of ecosystem sustainable development.

In order to approach my primary inquiry with a meaningful and interesting perspective, I have followed a research design organized into a three-phased process. Embedded in each phase of the research design, there are three subsequential research questions that will assist in building the argument and perspective to answer my primary inquiry by using a mixed methods methodology. The research questions are addressed through an open mixed methods approach (Tashakkori & Creswell, 2007; Teddlie & Tashakkori, 2006) composed of three complementary studies that are fit to create a multi-faceted understanding of this phenomenon. A mixed methods approach is appropriate, "…for the analysis of different facets of a phenomenon, yielding an enriched, elaborated understanding of that phenomenon" (Quinney, Dwyer, & Chapman, 2016: 4), and is particularly useful when the phenomenon being studied is relatively unexplored (Choi, Morgan, & Burnett, 1998). As noted by Greene, Caracelli, and Graham (1989),  a richer understanding of a poorly understood phenomenon can emerge through a sequence of studies that provide complementarity (i.e., each of the studies focuses on a different aspect of the problem), expansion (i.e., each study extends the scope of inquiry), and facilitation (i.e., the findings from one study inform the design of the subsequent studies.

My mixed methods approach combines qualitative and quantitative studies, with the nature of the research questions and data requirements guiding their sequencing, relative emphasis, and points of integration (Johnson & Onwuegbuzie, 2004). However, certain phenomena may require an alternate mixed methods design approach (Teddlie & Tashakkori, 2006). The QualQuant division and related approaches form the foundation of behavioral science research. Studying a meta-theoretical framing of ecosystem sustainable development integrated with the nested meta-organization, macromanagement, and dynamic capabilities frameworks seeks to understand emerging concerns and problem domains by building, applying, and studying the effects of an integrated systems management process for sustainable development in city ecosystem. The meta-theory of ecosystem sustainable development framework is a valid and complementary paradigm for exploring the internal-external phenomena of ecosystem management and dynamic changes they yield. The meta-theory of ecosystem sustainable development framework is valid for this research because it allows us to examine the ecosystem and its characteristics with the implementing meta-organizations' primary function of transforming a city ecosystem towards sustainable and flourishing form identities. The unit theories of macromanagement, meta-organization, and dynamic capabilities are complementaries that support the meta-theory of ecosystem sustainable development and builds the model of an integrative ecosystem management.

**Research Questions**

The primary research question of the book is:

*What are integrative ecosystem management approaches, and how are they instrumental in the development of flourishing cities as sustainable ecosystems (CASE)?*

The three studies that comprise this open mixed methods book respectively use: (1) Qualitative grounded theory development (Charmaz, 2014; Strauss & Corbin, 1997) with 21 sustainability managers, business and community leaders, activists and volunteers providing the lived experiences of sustainable development initiatives; (2) Quantitative Structural Equation Modeling SEM study (Bagozzi & Yi, 2012) operationalizing Appreciative Inquiry AI with the whole systems strengths-based process of the SOAR frameworks on innovative capacity and resilience (RES) (Stavros, Cole 2018; Bec, 2019; Scott & Bruce 1994; George & Zhou, 2001); and (3) fuzzy-set Qualitative Comparative Analysis (fsQCA) study to explore the role of a combination of factors in predicting improvement or unimprovement of city SDG total score based on the annual UN SDG 17 City SDG reporting (Lynch, 2019; Sachs et al., 2019; Teece, 2018b; Weymouth & Hartz-Karp, 2018). The fsQCA results in both in-depth case knowledge and generalizable results (Ragin, 2014) by using set theory and Boolean algebra to analyze impacts of combinations of factors (i.e., pathways) that lead to outcomes of interest (Rihoux & De Meur, 2009; Rihoux & Ragin, 2008). Figure 7 summarizes the research design, and Table 5 shows the research questions, thematic constructs and theory, and each unique study research design.

## Figure 7: Research Agenda Design by Process Phases

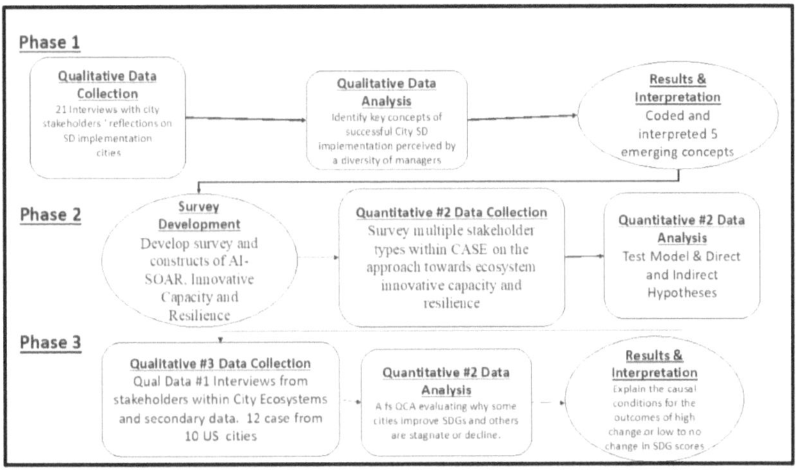


## Table 5: Integration of Research Questions, Theory, and Mixed Methods Research Design

| Research Question | Theory | Research Design |
|---|---|---|
| **RQ1. What are the factors associated with achieving successful sustainable development in city ecosystems? How are practitioners and managers approaching city level sustainable development currently?** | **Meta-organization Theory**<br>Multi-stakeholder Process<br>Developing Capacities<br>Adaptive Learning<br>Collaborative Integration | Qualitative study (Grounded Theory method) exploring the antecedents to city level sustainable development goal performance within the frameworks of capacity building and multi -stakeholder processes. |
| **RQ2. What components of Appreciative Inquiry and SOAR Framework leads to increases in innovative capacity and resilience in city ecosystems, which also influences the transformational development towards cities as sustainable ecosystem?** | **Macromanagement Theory**:<br>Prospection and Positive<br>**Meta-organization Theory:**<br>Appreciative Inquiry<br>Strengths, Opportunities, Aspiration, & Results (SOAR)<br>Resilience<br>Innovative Capacity | Quantitative study to test SEM model of the whole system approach in CASE by measuring the effects of Appreciative Inquiry/SOAR framework towards resilience and creative performance. |
| **RQ3. What combination of factors predict High or Low performance of City-level SDG achievement?** | **Meta-Organization Theory**<br>(governance device)<br>Capacity Integration<br>**Dynamic Capabilities**<br>Resource Orchestration<br>Adaptive Learning | Fuzzy-set Qualitative Comparative Analysis to why some city ecosystems improve SDG total scores while others stagnate/decline. |

To analyze my primary research question, I developed a three-phased process building from the frameworks of my key theories to explore supporting evidence to answer my inquiries. Figure 7 reveals the process flow of my research agenda and describes the methodological approaches I performed for data collection, mode of analysis, data results, and interpretation of various dimensions of developing cities as sustainable ecosystem CASE. In each research phase, a research question inquired about a specific phenomenon to analyze and confirm by empirical evidence to answer how multi-stakeholders can approach sustainable development in cities. Research questions, framing theories, and research design of each study can be viewed in Table 5.

In my primary research question, I asked to identify integrative ecosystem management approaches and explore how they are instrumental in the development of flourishing cities as sustainable ecosystems. To provide supporting evidence for this inquiry, I employed a mixed methods research design consisting of my three studies with unique research questions. The mixed methods approach is needed to answer my inquiry because one study may limit me to a reductive conclusion that sought to explain the complex phenomena at the scale of an ecosystem. Moreover, my mixed methods approach was designed for my three subsequent research questions to support the answer to the primary question. In Study 1, I asked what factors are associated with achieving sustainable development. The findings in research question one will support the primary inquiry by providing the key constructs that emerge as the building blocks for integrative ecosystem management approaches to ecosystem sustainable development such as a meta-organization, multi-stakeholder process, capacity building, and adaptive learning. Research question two seeks to confirm and validate that the properties of Appreciative

Inquiry and the SOAR framework as a platform positively impacts innovative capacity and ecosystem resiliency. This question was designed to support the primary research by providing a conceptual model that operationalizes the components of integrative ecosystems management: macromanagement, meta-organization, and dynamic capabilities; and supports the theoretic lens in which they are framed. Research question two also provides confirmation and validation that the combination of these three building blocks as an integrative ecosystem management approach is effective in supporting sustainable innovation and city resiliency development, which has positive influences on the evolutionary fitness of the ecosystem. Research question three asked what combination of factors led to high sustainable development performance or low sustainable development performance. The findings from this question will support the primary question, as they will reveal that high ecosystem performance towards achieving sustainable cities can be predicted by deeper integration of systems and resources, meta-organization, building capacities, cities' stakeholder awareness and attitudes, and led by powerful asset orchestration just to name a few factors. Research question three and its findings support the primary question in two forms: (1) It focuses the attention on the whole city ecosystem and acknowledges its underpinning systems dynamics, and (2) It examines the degree of integration of ecosystem management approaches, or lack thereof, and how they collectively contribute to the performance of the city ecosystem.

As presented in Figure 7 and Table 5, Phase 1 inquiries ask *what* emerging factors lead to the successful implementation of sustainable development projects in cities. A second research question was added in this phase, asking *how* management and practitioner (multi-stakeholders) are approaching sustainable development in cities. A qualitative analysis utilizing the Grounded Theory method was how I approached the research in Study 1. I explored what thematic antecedents emerged from interviews with various sets of stakeholders in the practice of city-level sustainable development initiatives. I collected 21 interviews with city stakeholders giving their lived experiences by reflecting on examples of sustainable development implementation in four cities. The purpose of this qualitative research was to discover how various types of stakeholders view successful sustainable development implementation within the boundaries of a city ecosystem. Using grounded theory, my research shares critical insights into the views and decision-making processes that stakeholders in the communities of Chicago, IL, Memphis, TN Charlotte, NC, and Cleveland, OH govern, implement, and manage sustainability projects. Lived experiences of such projects are elicited from sustainability managers, local NGO executives in sustainability, state, county, municipal government sustainability agency officers, university professors from a variety of disciplines, and volunteers that execute sustainability initiatives.

To analysis my data I used NVivo to perform the 3-stage coding process identifying key thematic concepts that emerged as factors of city sustainable development implementation success. As a result, five constructs emerged: meta-organization, multi-stakeholder processes (MSP), capacity building, adaptive learning, and collaboration. These constructs advanced us into a motivational inquiry of how these constructs can be conceptualized and measured in a causal logic. This progressed me towards a survey development research design by quantitative analysis in Phase 2.

Phase 2 was a quantitative study utilizing structural equation modeling (SEM) to measure and test Appreciate Inquiry platforms as a meta-organizing and whole systems management approach to foster innovation and resilience in cities. In this empirical inquiry, I asked *whether* an integrative macromanagement approach increases innovative and resiliency capacities. I developed a survey instrument, then collected quantitative data from multiple stakeholders that participated in sustainable development and innovation in cities. I then analyzed my survey data, following the testing of effects on the structural model with my direct and indirect hypotheses. As a result, my hypotheses were supported with statistical significance ($p<0.05$). The theoretic background in this study applied meta-organization and macromanagement theories to help frame the integrative systems management approach of operating the Appreciative Inquiry and SOAR process leads to more sustainable innovation while building adaptive resiliency within the city ecosystem. The results and interpretations of Study 2 motivated me to attempt the discovery of what combination of developmental approach factors determine city-level sustainable development goal (SDGs) scores, which is my third and final phase.

Phase 3 is a fuzzy-set qualitative comparative analysis (fsQCA) study that explores configurations of causal conditions in city-level sustainable development that predicts total SDG score improvement or not. The underpinning motivation of why I used the fsQCA method in this study is to analyze and interpret how some city ecosystems improve total sustainable development goal scores, while other city ecosystems' scores stagnate or decline over a three-year period. My results show that combinations of *Multi-level Government Intervention/Support*; *Financial Capacity Building*; *High Citizen Climate, Awareness, Attitude, Culture*; *Leadership Asset Orchestration*; *and structuring a Meta-Organization Platform can* lead to high sustainable development performance. In contrast, combinations *Lack of Powerful Core Leadership Asset Orchestration, No Meta-*

*Organization Structure/Platform, Poor University Participation*, and *Multi-level Government Lack of Support/Intervening Government Barriers*; *Low and Unaddressed City Citizen Culture, Awareness & Attitude Towards SD*; *Poor Financial Capacity Building*; and No *Clear Local Adaption Plan* led to low-to-no sustainable development performance in cities.

The combination of Phases 1, 2, and 3 provides integrated empirical evidence as to the solutions to how stakeholders should approach the interdependent tasks of sustainable development in cities for the transformational goal towards ecosystem flourishing. I discuss in detail my interpretation of the synthesized findings and results in the integrated discussion section. Within the integrated discussion, I will reveal how the application of macromanagement, meta-organization, and dynamic capabilities are the key conceptual elements of integrative ecosystem management and why employing integrative ecosystem management processes, in practice, enhances ecosystem capabilities that transform cities-as-sustainable and cities-as-flourishing ecosystems.

In the next three sections, I unveil the three-phased research agenda as three unique studies. Chapter 5 is the qualitative study, Chapter 6 is the quantitative study, and Chapter 7 is the csQCA study. Each study will include an introduction, theoretical framing/literature review, methodology, and discussion sections. Afterward, I will deliberate towards the combined interpretation in the integrated discussion, then conclude with limitations and implications to scholarship and practice.

## CHAPTER 5: STUDY 1 - QUALITATIVE STUDY - META-ORGANIZING IN AMERICAN CITIES

**Introduction**

The historical patterns of approaching complex challenges by reductionists approaches to managing complex systems such cities have yielded marginal success towards achieving city-level sustainable development goals, specifically #9 Sustainable Cities and Communities and #11 Innovation, Industry, and Infrastructure. Beyond the direct human socio-politico-cultural dimensions of ecosystem assets and SDG indicators, I believe these two SDG goals represent an encompassing functioning of city ecosystem vitality. Innovations developed, local industries, city infrastructure, and the sustainability initiatives that maintain CASE can indicate how well cities are performing towards increasing some degree of social well-being, economic prosperity, and healthy ecosystems.

Recent research in sustainable development almost exclusively examines organizational and systems levels of what firms are doing to promote sustainability (Berke & Conroy, 2000; Jepson, 2004; Portney, 2003). However, there is considerably less research on how CASE or city initiatives are planned, developed, implemented, and managed, as well as what influences the complex organizational effectiveness of these sustainable city initiatives (Conroy, 2006; Jepson, 2004; Portney, 2011; Saha & Paterson, 2008). The multiple stakeholders involved in implementing sustainability initiatives include managers in city offices, municipal developers, government agencies, businesses, academic practitioners, non-government/profit (NGOs), volunteers, and citizens. Engaging stakeholders in environmentally responsible decision making is a key prerequisite for multiple stakeholders to assume a greater role in the design and

90

development processes (Kasemir, van Asselt, Durrenberger, & Jaeger, 1999; Ravetz, 1999). A critical aspect in this process is to enable stakeholders to not only interpret and make decisions based on expert assessments but also to appropriately involve the relevant parties in the assessment process (Thabrew, Wiek, & Ries, 2009).

Based on the findings, these sustainability practitioners identified specifically five factors that were essential to sustainable development towards flourishing initiatives in their respective cities successfully. Interviewees describe that building capacity, increasing collective collaborative efforts, continuous adaptive learning, and a multi-stakeholder approach with a meta-organization leads toward shared sustainability goal attainment and perceived successful implementation.

In addition to the emergence of these core conceptual factors that are the perceived antecedents to sustainable development goal achievement, I also observed that organizational structures, approaches, and processes were conventional, reductive methodologies to strategic change management and development. The organizational structuring was siloed, fragmented, and was not planned nor implemented and treated with the whole ecosystem in mind. Since the approach, planning, strategizing, and execution were usually linear by techniques, at best, most examples of successful implementation at the time of interviewing yielded modulated, siloed results that only impacted a few stakeholders. Examples of multi-stakeholder processes were identified by respondents as a predecessor to sustainable development, but what also emerged from the conversations was the lack of integrating and coordinating the diversity of assets and stakeholders in the CASE to engage in a future designing process of how most stakeholders want to live within their ecosystem settlements. What was also compelling was the three examples of "deficiency" thinking articulated by respondents. About 90% of my interviews mentioned a deficiency of capital was a barrier to implementation success. This finding was also echoed in other literature that linked the lack of financial capacity to sustainable development implementation failure (Saha, 2008; Wang, Hawkins, Lebredo, & Berman, 2012). Most respondents stated that a lack of financial capacity, disengagement of youth and community participation, and no direct integration and coordination of sustainable development initiatives targeted to affirm socio-politico-cultural assets. Since the interviewees were from cities represented in the study that had populations greater than one million, it elucidated that in the typical top-down approaches to complex challenges overlooked the ecosystems' greatest potential and foundational asset available for the design process, and that is the human element, more specifically, the human mind. This gave me the impression that the critical assets to stimulate the CASE vitality were not well thought-out nor fully appreciated at the decision-system and managerial levels, cascading its ineffectiveness or marginal effectiveness comprehensively for the benefit of the whole. In addition to the emergence of the antecedents to successful development implementation and using the UN Sustainable Development Goals (SDG) Sustainable city ranking system as an assessment guide, I raise the concern that underperforming city sustainable development rankings may be linked to how well the decision system (management approach, meta-governing structure, and city ecosystem resource orchestration) integrates a comprehensive plan, strategy, and implementation to appreciate and promote flourishing in the city ecosystem by sustainable innovation.

My research questions are as follows:

*What are the factors associated with achieving successful sustainable development in city ecosystems?*

*How are practitioners and managers approaching city-level sustainable development currently?*

**Literature Review**

Although there is a long way to go, the literature states that to some degree, most cities have had success in programs and or initiatives that promote sustainable development city goals (Saha, 2008; Satterthwaite, 2010). This literature review will provide an overview of four key concepts as their relationship to sustainable development success. The key concepts are (a) multi-stakeholder process, (b) building capacity, (c) collaboration, and (d) adaptive learning agenda. Since sustainable development success includes multiple stakeholders, a process flow that is comprehensive and inclusive must be followed that will allow for multiple pathways and non-linearity (Peter & Swilling, 2014). How well city managers and stakeholders build capacities in finance, technical, infrastructure, and support also determines a cities' sustainable development success. With whom and how well sustainability managers collaborate is also instrumental in sustainable development success. Along with build capacity and collaborating with multiple stakeholders, the success of a sustainable development initiative can be determined by how the stakeholders learn, adapt, and implement in their environment, which are also tasks synonymous with building CASE resilience.

The emerging sustainability movement at the local city level has prompted many scholars to focus on the extent to which defined sets of policies and activities pertaining to sustainability are being integrated into local planning and development efforts (Saha,

2008). Municipalities have taken the most action on policies to do with land development and land-use planning; the least action has been taken on economic development and alternative energy development policies.

To date, the most populous U.S. cities have established sustainability officers and sustainability departments within local governments. These agencies have inherited the tasks that were once unique to urban planning and development departments in municipalities. The multi-stakeholder process (MSP) is a value creation process or approach that is designed to foster inclusiveness and decentralize control through polycentric governance and whole systems integration for one or multiple goals that involve multiple stakeholders. In organizational sustainable development literature streams, examples of MSPs have been explored in meta-organizations (Ahrne & Brunsson, 2008; Berkowitz et al., 2017; Berkowitz et al., 2020; Gulati et al., 2012; Laurent, Garaudel, Schmidt, & Eynaud, 2020), multi-stakeholder partnerships (Bäckstrand, 2006; Hemmati, 2002), and integrated strategic systems approaches such as macromanagement which have been operationalized as Appreciative Inquiry platforms (Cooperrider, Whitney, & Stavros, 2008; Cooperrider & McQuaid, 2012; Gulati et al., 2016; Whitney et al., 2019). The stakeholders can be from various unrelated groups or all parties affected by a particular issue (Bäckstrand, 2006; Buuren, Boons, & Teisman, 2012; Glasbergen, 2002; Hemmati, 2002). The multi-stakeholder process balances the flow of adaptive learning and planning, collaborating management, building capacity, identifying conflicts and convergence, negotiating to alternative processes, then back to management learning, re-designing, and collaborating again, creating a circular flow. As cities start to develop and follow multi-stakeholder processes, building capacity for sustainable development is an integral first step that is crucial to multi-organizational effectiveness (Ingraham, Joyce, & Donahue, 2003; Rainey, 2009).

Organizational effectiveness literature suggests that given the political and socioeconomic environments of a government, a higher level of sustainability is a result of greater organizational capacity, which reflects the successful adoption of various strategies to acquire capacity (Berman & Wang, 2000; Ingraham et al., 2003; Rainey, 2009; Weber & Khademian, 2008). Berman and Wang (2000) concluded that capacity building involves three core conceptual foci: developing technical capacity and financial capacity as well as increasing managerial and leadership execution.

The first conceptual focus was technical capacity. Technical capacity was described as a performance measurement, and Berman and Wang (2000) defined it as "…the ability to develop performance goals and measures and to overcome such conceptual barriers as distinguishing outcomes from outputs" (p. 410). Other scholars later found that it is necessary to acquire technical savvy and expertise in sustainability from professional institutions, universities, other research communities, and private consultants (Lubell, Leach, & Sabatier, 2009b). Lubell et al. (2009b) posited that "Developing human capital through internal professionalization as well as establishing complementary ties to external technical resources is fundamental for credibility for strengthening social norms, and for institutionalizing change" (p. 258). But technical capacity should be understood beyond human resources, technical expertise, and capabilities. In the context of city capacity building, technical capacities can also include the fostering of innovation capabilities by reconfiguring knowledge and material resources that increase innovation outputs as products and services. And, this technical development is innovative capacity, and at the core of the city ecosystem transformation towards sustainable developmental outcomes (Aryee, Walumbwa, Zhou, & Hartnell, 2012; Capdevila, 2018; Lawson & Lorenz, 1999).

The second focus involved in capacity building was developing financial capacity. Financial capacity is an organization's ability to assemble financial resources to support its operations and missions (Wang et al., 2012). While arbitrarily throwing more resources at any issue is not the sole solution of sustainable development, it is critical to develop and institutionalize funding mechanisms and explore financial resources such as grants, vouchers, loan guarantees, and taxes for sustainable development initiatives (Lester, 2002). In contrast, some scholars have studied the effects of capacity deficiencies in building technical and financial capacity (Bingham, Nabatchi, & O'Leary, 2005; O'Leary, Fiorino, Durant, & Weiland, 1999; Saha, 2008; Weidner, 2002; Werhane, Hartman, Archer, Englehardt, & Pritchard, 2013). These deficiencies were linked to everything from policy and implementation failures to low participation from political leaders and citizens (Saha, 2008; Weidner & Jänicke, 2002; Werhane et al., 2013).

The third focus involved in capacity building was developing managerial execution (capacity). Managerial execution (or capacity) reflects an organization's ability to develop sustainability goals and principles, incorporate those goals and principles into the strategic planning process and operations, and monitor and assess the achievement of those goals (Wang et al., 2012). Implementation in sustainability can be smoothed by having permanent institutional arrangements, such as designated individuals or offices in sustainability. As these offices increase human capital by adding personnel and managerial responsibilities, the managerial capacity increases and results in a concerted effort towards developing sustainability goals and comprehensive plans and increased

capabilities of implementing, monitoring, and evaluating performance in sustainable

development. Therefore, capacity (Dollery, Crase, & Grant, 2011) is associated with

organizational performance, and in the case of city-level sustainable development, it is

associated with multiple organizational performances collaborating together for the

common goals towards the CASE.

It is becoming common-place for macro-mangers to move away from traditional

hierarchal forms of management and leadership in sustainable development and evolve

towards a network or multi-level governance structure of collaboration (Armitage,

Berkes, & Doubleday, 2010; Bäckstrand, 2006; Berkowitz, 2018; Bouwen & Taillieu,

2004; Buuren et al., 2012; Gulati et al., 2012; Weber, Lovrich, & Gaffney, 2005) to share

resources in undertaking urban reconstruction projects, to build alliances between public

service agencies in regions and cities helping to coordinate services, and to avoid costly

overlap, inefficiencies, and redundancies. With various institutional logics, conflicts can

arise and may hinder performance; however, structuring a meta-organizational design for

governance (Berkowitz & Dumez, 2016; Gulati et al., 2012), and approaching

sustainability with a multi-stakeholder process (MSP), some conflicts may be expected,

but opening negotiations towards alternative processes and developing new capabilities

can create more collaborative opportunities for the whole ecosystem.

Building capacities and forging partnerships with adaptive learning processes can

develop new capabilities and innovation (Adner, 2017; Berkowitz, 2018; Chong, Habib,

Evangelopoulos, & Park, 2018; Linde et al., 2021; Teece, 2018b). As multiple

stakeholders collaborate, the collective practice of adaptive learning occurs, and this practice allows for knowledge sharing and generation, as well as innovation emergence on multiple levels spanning socio-economical and socio-ecological scopes and benefits (Armitage et al., 2010; Ostrom, 2010; Pahl-Wostl, 2009; Reed, Fraser, & Dougill, 2006). Adaptive learning practicing in creative collaboration is a crucial component of macro-managing in CASE-inclusive innovation that may be an organizational advantage utilized by CASE leadership (top-down) and peripheral stakeholders (bottom-up) for sustainable project implementation success. This co-creation of ideas from local and technical knowledge streams has been found to be essential in the process of inclusive innovation and in CASE resilience building (Tyre & Von Hippel, 1997).

**Research Design**

To discover the lived experiences of city planners, developers, government officials, local business and community leaders, and other peripherical stakeholders that dwell in cities, a qualitative research method was used based on grounded theory. Qualitative research is a tool used to observe social life in its natural habitat, and it enables researchers to produce a richer understanding of social phenomena (Yin, 1994). The use of grounded theory is appropriate when a researcher wants to make knowledge claims about how individuals understand reality, with grounded theory thus being integral to this particular study (Glaser & Strauss, 2009). These research methods provided the opportunity for participants to share their lived experiences through semi-structured interviews. Listening and understanding how these professionals discuss their roles in sustainable planning and implementation provided unique insight into their decision making and implementation process.

***Sample.*** In this context, the CASE is inclusive of the core municipality and surrounding suburbs. (e.g., Greater Cleveland Area, Chicagoland, Memphis Metro, and Charlotte Area) per the Metropolitan Statistical Area construct. The participants in this study consist of 21 subjects representing four American Metropolitan Statistical Areas: Charlotte, NC; Memphis, TN; Chicagoland, IL; and Cleveland Metro, OH, with sustainability leaders being interviewed in business (Clarke, Werestiuk, Schoffner, Gerard, Swan, Jackson et al., 2012), government (GOV), academia (ACA), non-governmental organizations (NGO), and volunteers/altruistic citizens (VOL). A fifth ad hoc group of the study was created to represent subjects from various cities. This group was designed to potentially capture additional insight from top-tier or mid-ranked cities that may not be captured in lagging ranked sample cities.

Twenty-one subjects were interviewed, with at least one participant of each subgroup in each of the targeted cities. These sustainability managers derive from various age groups, gender, backgrounds, and education, and are the driving forces of CASE development. The subgroups of each city included the lived experience from a government manager (planners, office of sustainability manager, public works manager), an NGO (non-profit, non-governmental managers), business manager (sustainability officer, developers, bankers, business managers), academic scholars (Interdisciplinary experts in sustainability), and volunteers. I have two tables that illustrate the demographics and sample characteristics. I refer you to Appendix B that shows participant demographics by interviewees in detail by gender, race, age educational background and level, industry, and position. The sample characteristics can be found in Appendix C. Appendix B displays the diverse and rich mix of interviewees specific to each city's sustainable development implementation. Appendix C shows the sampling characteristics across each of the studied city locations. Both tables can be found in Appendix.

***Data collection.*** The data for the research were collected from a total of 21 participants that were interviewed during an eight-month period from May to November 2017. Each interview lasted between 48 to 115 minutes. No incentive or compensation was awarded for participation in the research. Participants provided written consent that their interview be recorded and transcribed. A few of the interviews were conducted face-to-face in work and in non-work environments, such as in a home or another location where the privacy of the interviewee can be guaranteed. When it was not logistically possible to meet face-to-face, the interviews were conducted over the telephone or Skype.

As indicated earlier in my text, using grounded theory (Corbin & Strauss, 1990), interviews were open yet guided by a semi-structured interview protocol. The interviews began with a question about the respondents' personal and work histories. Demographic and other distinguishing data about the participants were significant in explaining their experiences. Open-ended questions were asked to elicit rich narratives about respondents' lived experiences. Audio recordings of each interview are transcribed into a Word document by the professional transcriptionist.

***Analysis.*** I began the data analysis with reading each interview transcript and listening to the audiotapes two times to ensure transcription accuracy, capture key words, and developing memos. Then I began coding, relying on a three-stage procedure (open, axial, and selective coding) (Strauss & Corbin, 1994). The first phase was open coding which involved creating tentative codes line-by-line reading for each transcript to identify "codable moments." In this stage, I distinguished words, phrases that might have potential significance. The concepts that emerged from each interview were reviewed and compared with other interviews until I could classify them under categories with specific properties, dimensions, and subcategories (Corbin & Strauss, 1990). From all the interviews, there were a total of 258 codable moments (Boyatzis, 1998) that emerged.

The second phase of coding was axial coding. According to Strauss and Corbin (1990), "axial coding is the process of relating categories to their subcategories." In this phase, I refined the categories, put new labels for them, and distinguished between categories and patterns. In this phase of coding, I refined and merged similar categories and subcategories and reduced the 258 codes to 20 labels.

The third coding phase was selective coding. In this phase, I identified the key constructs and findings of the study. I moved from a lower level of concept development (open coding) to a higher level of concept development, which termed categories or themes in the selective coding (Corbin & Strauss, 1990). After selective coding was performed, there were five themes that emerged for concept development and will be discussed later in the discussion section.

I composed memo writing through the data analysis in order to capture observations and insights, which lead to developing theoretical ideas. Theoretical sampling occupies a significant place in grounded theory. I sought pertinent data to develop the emerging theory and to elaborate and refine the categories by going back into the empirical world to collect more data about the properties of the categories (Charmaz, 2014). I refer to Appendix D, which illustrates this three-stage thematic coding process.

**Findings**

The data reveal that city-level sustainable development implementation is a

dynamic and complex process that requires multiple approaches to reach a level of

success for the initiative or project. Capacity building, together with learning agenda,

organizational effectiveness, and the overarching strategic approaches to sustainable

development, are crucial components that may have led to the successful implementation

of sustainable development projects in the four cities studied. The five key findings

below detail the sustainability managers' practices that emerged as themes leading to an

increase in the city's chances of successful sustainable implementation.

**Finding 1: Sustainability managers that continuously build capacity internally and externally have a greater probability of implementing successful sustainability projects.**

Ninety-five percent of the respondents shared their lived experiences in managing

the implementation of their sustainability projects as the process of building capacities.

According to the literature, capacity building involves the development of technical and

financial support for the project, collaborating with multiple stakeholders, along with the

dissemination of knowledge to all stakeholders (Wang et al., 2012). The study reveals

that the financial support and fiscal responsibility of a sustainability project is a crucial

component of whether a project is implemented successfully. Three of the study

participants spoke directly to this point:

> We created this program called the Energy Roundtable, where we partnered
> with UNC [city] and their students and professors. They would go into a
> building and do a free audit, level one, level two audit of the building. The
> reason we designed this is we heard from so many buildings that they were
> getting bombarded with these companies around energy efficiency, and they
> just didn't know ... They didn't have a way to gauge which was the right one
> for their building. ...Partnering with UNC, seeing that these students would
> go in, they'd give an unbiased opinion, which was very helpful to the

buildings to be able to say, "Okay, I need to do a whole lighting retrofit or I need to change my controls." It was hugely popular for the students because they actually get hands-on training on how buildings work and dealing with actually working people. Then the second piece of that, they created a two-day course for property engineers to come in and learn how to manipulate their systems of their buildings better, just kind of understand their systems so that they can become not just a property engineer, but almost like a property IT person to understand the complexity of the systems and so that was so hugely popular that we were able to secure a $500,000 deal, Department of Energy Grant, to expand that program from our 61 buildings to 200 buildings. — CHARNGO2

Almost all locally driven. We do apply for a fair amount of grants and get some of them from the state and federal level. There's Ohio EPA grants, for an example, from the state. Then at the federal level, the Great Lakes Restoration Initiative, so we've gotten a few grants, for example, through that. Then there's other foundation-based grants that we get as well. So, we're going to get up in our Climate Action Plan over the next year. The main grantor is a consortium of foundations under the name of Partners for Places. It's a wide variety of sources, but almost all the projects themselves in the programs are locally driven. — CLEGOV1

.... was kind of foundational in leveraging the head funding that helped implement the plan or implement the planning process of the mid-south regional green plan, so to put that document together they leveraged funding and it came from the fact that we had this foundational piece called the sustainable Shelby plan that has 130 directives, I think, written action items that we're trying to achieve in the mid-south regional green print. Now in its implementation phase starts to address some of those things that were originally developed through the sustainable Shelby plan. So, this idea of kind of building on and dove tailing initiative to initiative and growing our capacity to do green things and create success stories is really critical and I think that from the beginning of the office's foundation to today there's been a lot of really good groundwork laid for us to work with. — MEMACA1

It was interesting to find that most local governments fund the wages of their sustainability offices and small amounts of financial resources to engage stakeholders, but many of the projects are funded and executed from and by external sources. This leaves managers in sustainability offices dependent on resources from outside sources to manage and execute sustainability initiatives. Partnering with an organization that acts as a fiscal agent for city projects enhances the implementation success and flow of financial support for other sustainability projects. An example is given here from one respondent:

> ...we ended up with three years of funding. Half a million in year one, a million in year two, and three hundred forty in year three. Then it went to, the fourth year we were only able to raise forty thousand so we actually ended up one of the two funders had a requirement that the funding go to help food access for seniors, so we ended up because we only had forty thousand contacting our local public housing authority and saying do you have any senior buildings that don't have community gardens and can we help make that happen. — CLENGO1

Here, the NGO is collaborating with the Office of Sustainability acting as a fiscal agent and adding capacity for several projects. The capacity is expanded because the sustainability manager is not executing the project alone, but gaining technical know-how, increasing capabilities, and sharing responsibilities in securing and distributing funds for multiple projects at the time.

In contrast, since most local governments and nonprofit organizations rely on federal, state, and other financial support, the lack of building financial capacity appeared to be one of the greatest obstacles to implementing some sustainability projects. A good majority of the respondents shared this experience in their interviews:

> We rely heavily on federal funds for a lot of the work that we do, particularly around housing and sustainability......We are definitely focused on improving where we stand on a lot of these things, but funding is always gonna be a challenge. — MEMGOV2

> We also use funding, the Community Development Block Grant funding, CDBG, which is administered through HUD, that's the most flexible funding source we have. We get about six million dollars a year. That was proposed to be zeroed out as well. As well as HOME funds, which we receive about two and a half million annually, and that's used to support housing redevelopment, whether it's single or multi-family, and that was proposed to be zeroed out. So, we definitely have challenges from that front. So, our goal is to identify ways that we can control our own destiny, so we're currently exploring a housing trust fund, something that we can tie to a local

tax or fee that we will have every year that we can rely on to pay for improvements to housing in particular. — MEMGOV2

Yeah. I would say the only objection among city leaders or county leaders was that they needed to get the money from somewhere, and there was external fundraising that was done. I believe it was almost entirely paid for, and I actually don't know for sure, so don't quote me, but I do think that the majority of the funding came from private fundraising. And by the way, I did not say earlier that the piece of land where the Greenline is built is an abandoned railroad track right of way. — MEMNGO1

So, we ended up doing two community gardens at the senior facility in the fourth year because there was just a lot of funder fatigue, and we could not raise additional dollars. — CLENGO1

The total pot of money for this is, even the potential money, is too small right now. It cost more. Some things that really require a lot of investment, some things like trees, for instance. It's going to require millions and millions of dollars to grow a canopy back, and it's just not there right now. It's a lot of work over the coming years trying to pull that together." — CLEGOV1

Although the building of financial support can be a stumbling block for some cities, other cities take an approach to turn this challenge into opportunities to engage the private sector and build capacity through other means such as gaining technical know-how and expertise to implement a project to success. These public-private partnerships have emerged as a viable approach to successful city-level implementation and have been another catalyst for successful city-level sustainability implementation, especially when the private entity has an interest in the particular industry. Three respondents spoke directly to this point:

We were looking for something to do with [energy company], and we were looking as an organization for a project that would be city changing; that would be replicable taken from Uptown [city name] to another city or to another neighborhood in [city name], another place in [city name]. — CHARNGO1

But instead, we said, 'You know what, let's think about that. Let's see like what options there are out there,' and we reached out to a for-profit corporation in real estate to donate that space. So, I think that's one example of sustainability in the sense that you're not necessarily trying to generate something new, you're just tying in different key players that may be able to share resources, and you know, we all play in the same space. So instead of trying to use money to solve the issue, what is already out there and utilizing those, you know, resources for the benefit of that non-profit. — CLEBIZ1

Sure. It originally was started through [utility company] as a Clinton Global Initiative to start at 501c3. There were a lot of entities in [city], the Mayor's office, the Downtown Association who got behind this, and they really wanted to create a sustainability initiative for our uptown area, and they wanted to focus on energy efficiency. The way it came about was pretty cool in that [utility company] would put a lot of funding behind it and their strength behind creating a program that they could pilot through Envision and ultimately commercialize, but around the behavior side of energy efficiency. — CHARNGO2

All four cities had respondents that explained the importance of capacity building,

both physical assets and economic assets (financial) forms, as a crucial component that

drives the sustainable project towards success.

### Finding 2. All Sustainability managers are consistently learning through experiences, professional networks, and conferences to increase their chances of implementation success.

According to the participants' responses, there is a high level of learning that

occurs before, during, and after the implementation processes. The first observation of all

respondents was their educational level. All participants in the study had at least a college

degree, with many of the participants carrying multiple and/or advanced degrees. In all

four cities, two professional organizations repeatedly were acknowledged as sources for

learning and knowledge sharing for managing sustainable projects (43% of respondents);

they were the APA (American Planning Association) and the USDN (Urban

Sustainability Director's Network). The USDN was acknowledged as the central network

that most U.S. city sustainability managers participate, gain best practices, learn about

106

grant opportunities, and network with other sustainability managers to successfully drive

sustainability in their municipalities.

Most of the respondents shared examples of less than successful implementations

of sustainability projects, but a few respondents explained that they used that example to

learn how to proceed in the future with other projects (33% of respondents). One

manager, in particular, gave an example of how an energy waste reduction project in the

City Center that was considered a successful implementation was replicated using the

same tactics on a water waste reduction project in the Center City, and they learned that

the approach did not translate into success due to the selected sample of downtown

buildings that were recently built and already efficient in water waste management. The

participant responds below:

> [City] is such a new city, so most of our buildings have been built, you
> know, the big commercial buildings we've worked with have been built after
> the '80s. They're just much more water efficient, and the payback isn't
> there….
> When we started looking at water, we decided there's not a lot of savings
> in commercial buildings, so pretty efficient as they already are. What could
> we make a bigger impact on? We began working with the [city] Water
> Department to look at a scenario, a data scenario tool that could start by
> helping them identify nutrient sources into the water and then expand them
> into more water efficiencies. — CHARNGO2

Other managers learn from their own experiences, from NGOs, and by

replicating other city's successful implementation (86% of respondents). The

following quotes were captured from respondents that spoke to these findings:

> Really, just structural issues too on things like waste. It's just not as cost-
> effective, to do recycling here and composting as other cities. The amount
> of cost to dump in a landfill here or landfill tipping fee is, generally, much
> cheaper than most of the other country. Also, what we charge residents to
> pick up waste is less, generally. The economics aren't quite as good as you
> would have in San Francisco or New York. — CLEGOV1

So, we ended up taking a trip to Indiana, me and the facilities person, and we looked at a lot of the things they were doing there, and their sustainability initiatives did not just focus on energy. They actually had recycling as a part, and they had tanks with fish in 'em that they would use to grow plants and a greenhouse, and the inmates were getting all kinds of certificates. So, we really started to see how sustainability could have much wider implications on creating jobs and really have an impact on reentry. — MEMGOV2

In addition to pursuing additional knowledge and learning, sustainability managers that implement successful sustainability projects create knowledge as well, then share best practices with their network to create more synergies across the sustainability spectrum. Not only does this affect the potential success of future projects implemented by the managers in their municipalities, but it also creates more opportunities to expand sustainable practices regionally.

**Finding 3. Sustainable development projects that take a polycentric governance approach to management and have the right mix of leadership with working personnel tend to have successful implementation of initiatives.**

Good leadership is an essential factor for any successful endeavor, and those that work towards a shared vision or common goal are just as important when it comes to organizational effectiveness. This is no different for sustainability managers, mayors, governors, and citizens. Twenty-nine percent of respondents gave examples of good leadership in their sustainable development success. Seventy-three percent of the respondents gave examples of good collaboration efforts with external organizations as well as with key internal personnel. What was interesting that emerged from the interviews was the explicit mentioning of working with other people, departments, and organizations; and how the success of that common goal was linked to the interconnectedness of others. Three participants spoke to this:

…you need to have the right people to help you lead. From my role, I think the days have really come and gone where I can be a direct leader in that space. My job at my level is really to keep my senior leadership team focused on being able to deliver that high-quality service, which includes my sustainability director. Whenever you're talking high-quality service, it's always in terms of the end user. So, finding the right people. — CHIACA1

….and the challenge of siloed departments working more cross-functionally than ineffectively working in separated silos doing the same work. — CHINGO

…that's part of what it takes; it takes people who … people who really care about it and are willing to put their personal name on the line. You know, a lot of times doing sustainable projects for a city, sometimes it takes putting your personal brand on the line. And it takes courage. It takes real courage. — CHARNGO1

It is important to point out that two of the three quotes were directly from the

leaders of a nonprofit organization and an institution, which were included in the 73% of

respondents aforementioned. When leadership resonates positivity that positivity

transfers to others. Also, when leaders have the control of hiring or collaborating with

good people that are interested, engaged, and skilled at their task, positive results resonate

and can lead to a higher level of performance towards the shared goal. On the other side

of the coin, employees of leaders share their experiences of what good leadership is

within private and public organizations. The following quotes capture this point:

Yeah, our CEO is … who is actually a CASE MBA grad himself, and he's on the board with a lot of practices revolving around sustainability with CASE. He has championed this as our purpose of an organization, where we are creating an environment beyond going after revenue goals. We have a purpose as an organization to give back to the community we … and the employees and people that all work together to make this thing possible. So yeah, we … that's part of not just how we're managed, but that's just part of who we are as a company, that's our purpose, it's written above our mission statement for clients. — CLEBIZ2

And so, we're starting to see that through our communications both with the city of Evanston and the city of Chicago, and it's great to see how the mayor even, he basically replicated the EPA website so that we can keep those resources available to us at our state level. So, I, it's incredible to see how much reinforcement, and how it's still continuing to move in the right direction. And there's now more urgency, if anything, to get it done. — CHIACA2

...the mayor wanted to engage the whole community, and now they're going to co-create with sustainability means for everyone.
... I think certainly community-driven in a sense that there was a demand for city leadership in those, but also a way for people to work together. A big factor was the mayor went to a, um, was like a UN convening here, United Nations, and he saw that it was being facilitated by a couple professors. — CLEGOV1

Leading by example, I think, is a really great example, too, but communicating our successes which is the county mayor is really big on that, is something that we're striving to do but I don't think we've always been able to communicate successes to the people well.
And she was telling me that the Mayor was so excited about the article that he's ready to do more, so you know, all of those things really help spread the message on the value of sustainability. — MEMGOV2

Good leadership can also resonate within the local political system as well, while leaders

that are perceived poorly can be a key reason why a sustainability project is perceived as

less than successful.

In some of the interviews, some sustainability managers expressed the need for

better leadership in elected officials. Here is a sample:

...it's been a challenge, I think, to get some leadership and elected officials to support moving ahead on making sure government is on the leading edge of sustainability issues. I wouldn't say that's across the board. There's been some great done stuff in government buildings, specifically county corrections who's done a lot of great, green sustainable initiatives. But I think, as a whole, there could be a lot more to be done. So that's been a challenge. — MEMACA1

During other interviews, participants shared detailed experiences of poor

leadership of some elected officials. Some of the details include internal conflicts,

political transitions, and personnel changes within all organizations that collaborate on

sustainable development projects. Sometimes when these events occur, it can slow down or even halt the process of implementing the sustainable development initiative and make

the project perceived as unsuccessful. Examples of interview quotes from a few

participants of the aforementioned events are below:

> One of our big goals is to expand bike infrastructure in the city and increase cycling and improve health outcomes and things like that, but some council people just, for whatever reason, hate bikes, so they will fight tooth and nail to have any bike lanes in their ward. It's one example. — CLEGOV1

> This leader did not believe in climate change from a religious perspective, and when you don't have buy-in from the leadership above, it makes it hard to initiate or get pull through for projects identified to bring sustainable solutions to the organization. — CHINGO1

> The ability for us to review a project and approve a project based on its business case occasionally are thwarted by political action. Where the legislature or the governor really wants a project to go ahead even though it's not economic in a sense and kind of pushes us to approve the project. Then we'll have projects that don't, aren't sustainable and then we get a lot more grief from someone else. Sometimes that's the community that does that and sometimes it's the state. It's politics more than anything else. — ADHOCGOV

> But there was federal funding called HHF funding, and the rules were that the person who got the HHF funding for demolition had to hold the land for three years or transfer it to an adjacent neighbor as a side lot, and what our county land bank did was they opted in many cases to hold the land for three years and then they launched their own version of the in-land reimagining program. That got really complicated because the city, even though it was a federal rule that they hold the land, there were issues because at times they were making decisions on who got the land, and the city that was supposed to be a process that they didn't do that without city approval because the city wanted to be the final determinate, they didn't want the county land bank deciding. There was some tension there. They also were very interested in planting trees on vacant land. The city of {city) said it was too hard to mow the grass if there were trees, they didn't want the trees, so there's just been a lot of frustration, tension, non-consensus between the county land bank and the city land bank, which ended up being just a problem for the second thing. Which was the second thing being we had a study done on vacant land, and it was initially intended to be an assessment of the seven agencies who had all proliferated some kind of vacant land program. — CLENGO1

> Well, one of the big challenges is turnover, staff turnover. You're dealing with one of the property managers or property engineers, and it gets bought

by another management company or the building gets sold, so you're
constantly ... Trying to keep track of 61 property engineers, it was
unbelievably challenging. As soon as you schedule something in a building
and you start implementing it, that property manager was gone, and a new
person would come in. I think that was the biggest barrier we had on that
program. — CHARNGO1

**Finding 4. Multi-stakeholder processes are essential and necessary for collaborating, planning, and strategizing of managers to implement sustainability projects, and as of yet, there is no universal or "one size fits all" set of solutions that works best for every city.**

The majority of the respondents gave examples of successful sustainable project

implementation that dealt with city-level issues of affordable housing, land use, economic

development, and energy reduction. Most are issues city planners and urban designers

have been attempting to bring solutions and balance for decades. Appendix A includes

shared lived experiences and the type of projects successfully and unsuccessfully

implemented according to sustainability managers' responses.

The table in Appendix A shows that the majority of projects mentioned by the

participants, both successful and unsuccessful, dealt with some of the issues traditional

urban planning and development attempted to manage. These issues were inherited by

office of sustainability departments across cities; and sustainability managers now have

to collaborate with more stakeholders and facilitate more innovative ideas through their

cities to implement better sustainability outcomes. This possibly describes multi-

stakeholder processes. Multi-stakeholder processes (MSP) are becoming popular when

organizations, communities, and cities are designing, implementing, and monitoring their

sustainability goals. In (MSP) there are concerted efforts by many different stakeholders

approaching a multitude of issues and goals at the same time. Although there were no respondents that gave a successful implementation example following multi-stakeholder processes, a couple of them described the process:

> In my mind, more importantly, when I can line those three, four, five, or six problems up in just the right way from a puzzle perspective that I can drive one solution through all of them, I've linked those three, four, or five, six issues together through the conduit of that one solution, and they stay linked together so that you don't have to worry about what you're doing ending up in a vacuum. Once you've linked those things together, you've made that connection, then other connections start to appear. — CHIGOV1

> When you talk about groups like [Children's Hospital], they do a lot of work with kids that have asthma, and one of the things that we found is that the greatest trigger for asthma is their home environment. So how can you go into that home, and you make remediation, whether it's removing mold and lead and improving indoor air quality by changing out HVAC systems, which also ties into weatherization, how can we go into that home, make it more healthier, make it more energy efficient where you maximize the impact? So then, if you're reducing health costs, then we have different groups that we can go to try to get revenue for it. So, if we're reducing how much it costs to them to see a patient that's suffering with chronic asthma, then there's an interest from insurance companies in investing in those interventions. So, I think that that is going to be significant for this community once we identify the structure for how we, one, identify those patients and where we can have the greatest impact, but then, two, going in and actually doing that remediation work. I think that is a significant opportunity in this community. — MEMGOV2

Linking seemingly unrelated sustainability indicators such as healthcare cost, clean air, and home weatherization; collaborating with multiple organizations and stakeholders; and simultaneously implementing several policies and programs are all components of multi-stakeholder processes. It is also essential to managers for building capacity and capabilities in reaching sustainable development goals.

**Finding 5. Based on the literature, one of the greatest opportunities towards city-level sustainability goals is to improve social equity outcomes. There were very few examples of successfully implemented social equity initiatives described by the respondents.**

Only 14% of respondents reported no lived experience examples of successfully implemented sustainability initiatives that directly approached social equity issues. However, there were examples of successfully implemented social equity initiatives mentioned from respondents that affected social equity issues indirectly through other programs:

> In our opening dialogue, as we started to get to know each other, I told you one of the things I'm proud about is Rebuilding Together, because it keeps people who have been living in those neighborhoods for 40, 50, 60 years in the neighborhoods another five, six, 10 years. I believe that anything that we can do to help stabilize our neighborhoods is a general improvement to our society. It helps reduce crime; it helps provide more equity into those neighborhoods. Those long-standing residents are the strength of our community. Anything that we can do in sustainability, in my mind, is a social equity or a social justice. Cause to diminish the difference between haves and have-nots is a very, very important and worthwhile cause. — CHIACA1

> It's also a way to enhance health in communities. So, there's so many diverse benefits. We had a strong focus on equity, which I think has continued to be at the top of the conversation when it comes to planning. — MEMGOV2

> Now we're looking at diversity inclusion being part of sustainability when that really wasn't a conversation five years ago. — CHIACA2

A lot of the indicators of social equity issues that sustainability managers face are associated with wicked problems cities have been dealing with for some time. This will be discussed further in the next section.

**Discussion**

Scholars and practitioners view sustainable urban development as the contemporary paradigm to address the challenges cities face and provide an opportunity to building a desirable urban future (Luck, 2007; Runhaar, Driessen, & Soer, 2009; Sachs et al., 2019; Yigitcanlar, O'connor, & Westerman, 2008). Yet research in urban sustainable development studies specific to U.S. cities is just beginning to emerge (Berke, Backhurst, Day, Ericksen, Laurian, Crawford et al., 2006; Portney & Cuttler, 2010; Saha & Paterson, 2008). One of the interesting findings in this study revealed that most offices of sustainability in the U.S. are nascent in form and have just recently completed drafts of sustainability comprehensive plans for the CASE. This particularly interesting because this newly formed agency within local governments acts both as a critical central node of the sustainability network and influential leaders involved in the planning, implementation, and evaluation of the many projects the cities adopt for sustainability. These professionals are a center point for collaboration, process management, research, and learning in the transition to sustainability in cities. Many of the sustainability departments in the U.S. evolved from urban planning and development or public works agencies within city governments, so it was not surprising to see a number of sustainability managers from government agencies in the study were planners, architects, and engineers. Urban planning research has developed for decades, but there is a scarcity of inquiry into the implementation process in planning, and even less in sustainable development implementation. However, after the interviews were coded, there were four themes that emerged: capacity building, learning, multi-stakeholder process, and collaboration. These themes may be strong factors that led the complex mix of stakeholders to successful sustainable development implementation.

*Capacity building.* Building capacity in sustainable development involves developing technical and financial support, increasing managerial execution, establishing sustainability goals, incorporating goals in the operations, developing  a supportive

infrastructure, and getting all stakeholders involved (Wang et al., 2012). The data reveals

that when city sustainability managers were able to expand and grow each of these

components of capacity to a degree, then the sustainability initiative implemented had a

greater chance of being a success. One example of a successful project included all of

these elements of capacity building. This city sustainability project had financial support

from a large utility company, got multiple stakeholders involved, including students, the

business community, and NGOs, then set out a timeline and goals. The examples given in

the findings reveal that from the nonprofit organizations that manage sustainability

perspective grants and charitable contributions are the lifeblood of their organization's

existence, and the continuous flow of funds builds financial capacity to successfully

implement the project and potentially other initiatives. In contrast, the data in the study

revealed that the ability to build financial capacity tended to be a top reason why some

sustainability projects were less than successful. In some examples, the project was

planned, but the managers could not get the proper funding or grant. In other examples,

managers found that while implementing some projects, they got to a point where the

efforts and the tasks were just not financially feasible to complete. In her 2009 survey

Saha (2009) came to this conclusion as well. She found that the lack of adequate funding,

elected official apathy, and lack of knowledgeable staff were the main three barriers cities

face when implementing sustainability initiatives. These three barriers speak to

deficiencies in capacity building: financial capacity, getting all stakeholders involved,

and developing supportive infrastructure, respectively. In this study, one example

specifically stated that they would be more successful implementing sustainability projects if they had the funding and staff capacity beyond the two people that currently were staffed. Two other barriers mentioned by Saha (2009) were the lack of public demand and acceptance, and opposition from the developer and business community; which also speak to deficiencies in capacity building as it relates to deficiencies in getting all stakeholders involved and building a supportive infrastructure. One difference found between Saha's study and this study was that at the time of Saha's survey, the City of Cleveland had no office of sustainability and only one sustainability manager working out of the utilities department. Eight years later, at the time of this study, the City of Cleveland had established an office of sustainability and has a staff of eight. This reveals that there is a fairly high level of activity happening around sustainability in the city and displays the city's effort in moving towards a CASE.

*Adaptive learning.* Examples of learning emerged from this study, and successfully implementing sustainable development initiatives in the city is a continuous learning process (Pahl-Wostl, 2009). Learning can also be a subcomponent of capacity building in the sense that learning builds trust, gains technical know-how, and allows for innovation to emerge (Armitage et al., 2010; Kay, Regier, Boyle, & Francis, 1999). This study had many examples of participatory learning by experiences (Roling & Wagemakers, 2000). Some sustainability managers found that learning from their professional network was a way of staying informed and building know-how (Davidson-Hunt, 2006). Other examples given found that learning from experience, specifically, sustainability projects that were less than successful was key to figuring out how to approach other projects to minimize failure, in which also follows the flow of multi-stakeholder processes. All in all, sustainable development is a continuous learning process, and managers have to stay current, engaged, and innovative to increase the likelihood of implementation success.

 ***Collaboration and integration.*** Support from the business and nonprofit communities is consistent with collaborative planning and collective action efforts that are germane to building social capital and to comprehensively tackling complex issues such as sustainability through the governance of public and private partnerships (Lubell, Feiock, & Handy, 2009a). I agree with their argument that more public-private collaborations are necessary for successful implementation of city-level projects and for cities to create the impact needed to become sustainable city ecosystems. Clark (2006) also explained in his work that public-private partnerships are emerging as a new approach cities have to manage shared resources, "The key to most progress is a new relationship with the private sector. Not just public-private partnerships, but a more advanced means to shared risks, costs, returns, and the stewardship of [orchestrating] assets" (pp. 10–11). From the data in this study, most of the examples given show that multiple collaborations have led to successful implementations of projects. In contrast Saha and Paterson (2008) argued, "As a result, often environmental advocates, developers, and decision makers butt head rather than collaborate to accomplish rational green infrastructure planning" (p. 34). This can indicate the lack of a macromanagement approach where everyone participating is involved in the process, or the absence of meta-governance structure where there is less than available understandings of clear shared purposes, responsibilities, and accountability of the future outcomes. The failure of macromanagement and meta-organization was also revealed in this study, most of the time leading to less than favorable sustainability outcomes or at worst, not implementing projects at all.

*Multi-stakeholder processes and approaches.* Since it has become apparent that neither government, business, nor social organizations alone will accomplish the transition towards cities as sustainable ecosystems, and multi-stakeholder processes and more inclusive programming to foster innovation and sustainable development in cities are emerging as the process for achieving sustainable city goals. MSPs that incorporate a macromanagement approach and meta-organizational design to develop dynamic capabilities forms an integrative systems management approach towards transforming CASE.

## Implications

In conclusion, it is a complex and difficult task to tease out all of the factors that make sustainable development projects successful and what makes them fail since the phenomena embed a multitude of complexities and emergent behaviors. Yet, various business and institutional leaders and civic stakeholders are integrating their efforts by pooling capabilities, knowledge, and expertise towards building a more sustainable ecosystem. Another challenge with whole systems analysis and conceptual approaches is the empirically designed methods for research in sustainable development. This is due to the level of complexity and continuous evolution and growth of all of the moving parts, factors, and components that influence the behaviors of groups, communities, and institutions collectively as one isomorphic field. However, some factors may possibly lead to better implementation successes. As cities move towards the goals of sustainability and further towards flourishing, city ecosystem management has to continue to orchestrate the building of capacities, create more collaborations and partnerships, adapt and learn when there are opportunities for configurational system changes, and approach sustainable development utilizing multi-stakeholder processes such as integrative ecosystem management. Although there would probably be failures in implementation, integrative systems management should expect that their efforts, adaptive learning capacities, and responsiveness to the ecosystem should lead towards positive outcomes.

As the meta-organizational structure and collective capacity building with multi-stakeholder processes emerge, I find that these concepts are aligned with the structure and processes inherent to complex adaptive systems; furthermore, they are aligned with integrative ecosystems management. As government leaders, business leaders, community leaders, NGOs, and academics work together; there me a link to successfully implemented sustainability projects by being more inclusive of community stakeholders following a multi-stakeholder process, encouraging participation, collaborative efforts, adaptive learning from one another, and building essential capitals and assets that are beneficial for all stakeholders. The more successfully implemented projects a city performs, the closer that city will get to balancing the scales of the three Es (Environment, Economic, and Equity) and reaching a healthy city as an ecosystem with social well-being, economic prosperity, and thriving ecological system flourishing. This should not only result in a better quality of life perception for the city's citizens but also move that city towards the goal of a CASE. Starting with city leadership capacity and the institutions that support them, or the orchestrating administration approach to management, this can be achieved by increasing human, social, cultural, political, financial, built, and natural vital assets of the flourishing city for most, if not all stakeholders dwelling in the city. Reconfiguring a system on the scale of city is complex and requires integrating approaches for change instead of problem solving in fragmented and reductionists methods. Encouraging stakeholder participation and finding creative strategies to build collaborative capacities is also important towards the transition, as classical top-down approaches have not been as inclusive as they can be. Additionally, there are opportunities for leaders to empower stakeholders and create elective affinities for citizens to get involved in multi-stakeholder processes earlier in the planning and designing stages of city development, and this has been found to yield positive outcomes towards flourishing (Glovis, Cole, & Stavros, 2014; McQuaid, 2019). Because city ecosystem change is a slower process, compared to individual or organizational change, it will take time to achieve certain SDG goals, and the key for managers and participants is to keep consistent engagement opportunities, develop a temporal cadence to facilitate engagement, collaboration, and idea generation that can materialize innovation, and support learning experiences by formal and informal methods.

The data from the interviews, the synthesized literature, and the five thematic concepts were considered in formulating a conceptual model and designing the structural equation modeling (SEM) for Study 2. To form a model using the constructs of meta-organization, multi-stakeholder process, adaptive learning, capacity building, and collaboration, my attention was drawn to Appreciative Inquiry (AI) platforms and its utility on the scale of cities. Researching the AI literature resulted in discovering a measurable construct for AI as an organizational system and the SOAR framework (SOAR) as the process that guides strategic planning and co-designing. Theoretically, AI-

121

SOAR platforms are structured to meta-organize, and the macromanagement approach is fixed within the AI design process towards achieving the shared goal. So, those are two of the thematic concepts I build from this study. Underpinning the AI-SOAR construct were survey items related to the properties and functions or components of AI-SOAR. These components included metrics of adaptive learning, collaboration, stakeholder needs, and generativity, along with several others that will form the AI-SOAR construct. Two of the thematic concepts from this study, collaboration and adaptive learning, will be embedded as a property of the AI-SOAR construct. A simple conceptual model was built to hypothesize the relationships between the construct of AI-SOAR and building ecosystem innovative capacity and resiliency constructs. Building innovative capacity on the scale of an ecosystem encompasses the capacity building of new artifacts and infrastructure, as well as technical, financial, managerial, social, and cultural capacity building. Therefore, the fifth thematic concept from this study, capacity building, will be adopted as the construct innovative capacity. All the thematic concepts from this study were incorporated within the conceptual model to test in the study of my second phase of the research agenda.

The second phase of this research agenda is a quantitative analysis conducted between 2019 and 2020, and based on SEM (Bagozzi & Yi, 2012). I extracted the thematic concepts from Study 1 and used them all, to some extent, in forming the conceptual model to test and validate in this study. Study 2 is theoretically approached within macromanagement and meta-organization frameworks conceptualizing, operationalizing, and measuring city ecosystem transformation towards a flourishing vitality. Staying consistent with the foundational conceptualization of complex adaptive

122

cycles of ecosystems, I highlight the isomorphic field of cities as sustainable ecosystems CASE along with the paradoxical ecosystem characteristics between innovative capacity and adaptive resiliency. I align the underpinning constructs of a whole-system multi-stakeholder process that combines collaborative organizational partnerships, builds adaptive capacities for innovation and resiliency by learning and seizing opportunities, leadership resource orchestration, positive and future-focused strategies towards positive implementation for socially and environmentally beneficial ecosystem outcomes. In doing so, I develop a hypothesized structural model and survey instrument that measures and tests Appreciative Inquiry (AI) with the strengths, opportunities, aspiration, results (SOAR) constructs' positive effects on innovative capacity and resilience in city ecosystems. I also hypothesize that interactions between innovation and resilience, displayed as dual mediation effects of AI on resilience RES mediated by innovative capacity IC and AI on innovative capacity IC and resilience RES, produce a flourishing ecosystem vitality. I hypothesize that this double mediating path effect interprets as the dynamic transformation towards a flourishing CASE.

Study 2 constructs were operationalized with a combination of latent constructs and sample demographic characteristics, adapting existing scales as summarized in the research design section. The survey instrument was slightly modified from the AI and SOAR survey instrument of Cole et al. (2019) to reflect items describing another focal person's behavior rather than their own behavior. Respondents rating someone else, such as their leader, or someone else whom the respondent is familiar, may perceive traits and competencies in others better than ratings the individual makes in him or herself (Libbrecht, Lievens, & Schollaert, 2010; Van der Zee, Thijs, & Schakel, 2002).

Additional constructs and measurement items in the measurement scale, such as IC, were developed from Zhou and George (2001) and Scott and Bruce (1994b), and the RES construct developed by (Bec et al., 2019). The subject population was developed through my past experiences in several professional organizations, and the professional network as a business consultant in sustainable development, leadership management, and innovation sectors. Items and constructs were validated according to standard procedures, including Q-sort, SPSS, and Mplus, and the measurement scale was pretested with preliminary surveying before inclusion into the research field (Anderson & Gerbing, 1988). Final results and interpretations were offered with outcomes of positive direct and indirect effects tested, implying a positive relationship between Appreciative Inquiry platforms as an operationalized process and ecosystem innovative capacity and resiliency. Moreover, the statistical significance of supported hypotheses validates my assumptions that meta-organization and macromanagement develop capabilities to forge sustainable innovation and resilience within the ecosystem, which in turn, supports a flourishing ecosystem vitality in cities. The overall structural model and causal logic validated and confirmed can be interpreted as integrative ecosystem orchestration for development towards flourishing CASE.

**CHAPTER 6: STUDY 2 - QUANTITATIVE: A MACROMANAGEMENT
SCALE DEVELOPMENT**

**Introduction**

Developing cities as sustainable and flourishing ecosystems is one of the greatest

societal challenges of our time because it will require complementary actions by

governments, academia, civil society, science, and business (Cooperrider & McQuaid,

2012; Sachs et al., 2019). Macromanagement has been viewed as a viable means for an

integrative systems management approach in achieving cities' sustainable development

goals along with societal flourishing. More specifically, Appreciative Inquiry as systems

management strategies and summits have been touted as a macromanagement approach

for whole systems, strengths-based approaches that are instrumental for macro-level

processing such as organizational systems, communities, and city ecosystem

(Cooperrider et al., 2008; McQuaid, 2019; Whitney & Trosten-Bloom, 2010).

Cooperrider and McQuaid (2012) describe Appreciative Inquiry summits as,

> …large group planning, designing, or implementation meeting that brings a 'whole systems ' of 300 to 1,000 or more internal and external stakeholders together in a concentrated way to work on a task of strategic, and especially creative, value [and] ...where everyone is engaged as a designer, across all relevant and resource-rich boundaries, to share leadership and take ownership for making the future of some big league opportunity successful. (p. 2)

However, empirically, macromanagement and Appreciative Inquiry (AI) platforms have

been described mostly in abstract concepts in frameworks of strengths-based, future, and

positive focused, while also being a distinct contrast to traditional micromanagement.

Existing within the literature streams, macromanagement and Appreciative

Inquiry are associated with systematic flourishing outcomes (McQuaid, 2019), bringing

out the best in human systems (Cooperrrider, 2012), are communication strategies for practitioner co-creation towards deeper meanings and identities of power through

conversations (Whitney, 2012), generating positive action for large-scale learning,

transformation, and social innovation (Whitney et al., 2019), and transforming urban

environments (Rosado, 2008; Sachs et al., 2019). However, macromanagement has been

criticized as being too broad of a concept to draw any meaning and poorly defined, and

the criticism is echoed by what is meant by Appreciative Inquiry. Moreover, critics of

Appreciative Inquiry have argued that it has limited perspectives focusing only on the

"positivity" that foster unrealistic views of human experiences (MacCoy, 2014), and the

process has tendencies to polarize "negative" and "positive" aspects of phenomena

(Bushe, 2012). But as the empirical evidence evolved over the last few decades, the

confusion and inconsistencies about AI have been somewhat resolved; especially, since

the emergence of the SOAR framework with the feature of "language" flipping problems

and weaknesses into opportunities (Cooperrrider, 2012; Stavros & Hinrichs, 2009).

The SOAR framework is a strategy formulation and planning framework that

allows organizations and systems to think, plan, and inspire implementation towards the

most preferred future (Stavros & Hinrichs, 2009). Strengths, opportunities, aspirations,

and results are elements that represent the acronym of the SOAR framework. Instead of

the analytical strategic framework established as strengths, weakness, opportunities, and

threats (SWOT), SOAR acknowledges a holistic dialogical perspective of present and

future timeframes with internalized and externalized scopes between human and their

environments for change management. In doing so, leaders and managers approach

change management with positive foci on strengths and opportunities instead of focusing

on weaknesses and threats in strategic analyses. The strengths-based, positive focus on change or value proposition opportunities motivates management towards the development of positive aspirational goals and inspires them to implement and act on achieving those goals for positive outcomes and results. This is a stark contrast from classical change management problem-solving by narrow reductive reasoning and approaching change as recursive responses to weaknesses and threats. As Cooperrrider (2012) argued, "No accumulation of simply fixing weakness and reacting to threats in the environment will transform a person, organization, nor system; only by amplifying strengths will they excel towards full-spectrum flourishing" (2012: Executive Summary p. ii-iii). The SOAR framework has been studied as an effective strategic process in planning urban tourism development (Khavarian-Garmsir & Zare, 2015), for encouraging collaborative capacities in teams (Cole et al., 2019), and leading change management for sustainable performance (Glovis et al., 2014). In combination with motivation, the SOAR framework has also been discovered to enable organizational flow for better performance outcomes in entrepreneurial endeavors (Stavros & Saint, 2010).

In this study, I combine AI with the SOAR framework with a conceptual understanding that AI is the "system" and SOAR is the "processor" of a macromanagement approach. Although the literature is robust about the effectiveness of change management associate with AI-SOAR as it relates to organizational or whole systems flourishing, there is an opportunity to propose a model of AI-SOAR effectiveness towards flourishing on the ecosystem level. To help leaders, business and institutional managers, and all other stakeholders in operationalizing a macromanagement approach to foster sustainable innovation and build resiliency, I offer a measurable model by conceptualizing, operationalizing, and measuring AI-SOAR effects on a city ecosystem. Operationalized, AI-SOAR as a platform would enable stakeholders and organizations to organize, make collective decisions, orchestrate resources, co-design, and integrate implementation for developing innovation while simultaneously building resilience for a flourishing city ecosystem. This is the first objective of this study. The second objective of this study is to identify an underpinning effect between innovation and resilience, which has been argued by Dale et al. (2010) as the adaptive interaction that grows and maintains the flourishing community or ecosystem. Thus, my research question is:

> *What characteristics of Appreciative Inquiry and the SOAR Framework lead to increases in innovative capacity and resilience in city ecosystems, which also influences the transformation towards cities as sustainable ecosystems?*

My proposed model was analyzed and tested using structural equation modeling to show that the macromanagement approach functioning as AI-SOAR platform summits results in positive performance and outcomes in ecosystem innovation and resiliency, which develops the city ecosystems' form towards a flourishing CASE. After presenting my results, I will deliberate on the findings and discuss the implications of extending the scholarship in the macromanagement and AI-SOAR literature; and offer utility of this instrument to management in practice.

**Theoretical Framework**

***Building innovative capacity and resilience by macro-management.*** Evolved from Meadows' (1999) systems management strategy, the macro-management approach extends decision-making capabilities from core leadership and is inclusive of internal, external, and future stakeholders pertinent to the ecosystem {Cooperrider, 2012 #648}. It is a multi-stakeholder process, defined as value creation processes or approaches designed to foster inclusiveness and decentralize control through polycentric governance for one or multiple goals that involve multiple stakeholders in a group, organization, city ecosystem, or ecosystem. The macro-management self-organizing system combines top-down dynamic capabilities, which are considered the strengths of organized systems with their technical competencies; and bottom-up dynamic capabilities brings distinctive strengths of local knowledge, inspiration, and aspiration. Collectively, the combination of both strengths has the potential of creating exponential innovative power and building resiliency within CASEs.

Beyond the property of integrating collective action, there are two other distinct properties of macromanagement that attribute to its uniqueness compared to other organizational systems approaches. First, embedded in macromanagement is a positive focus on the "whole" drawn by open communication and generative crowdsourcing of everything positive, strengths, and ideas of organizations and systems. For example, during planning and strategizing for system change, the knowledge sharing of information and idea generation would have a positive perspective, design, and conversations around certain meanings and identifications of resources, leadership capabilities, stakeholder needs, and opportunities for change. This means instead of strategic planning from problems and negative consequences that emerge, then approach them with reductionist strategies; conversations in strategic planning would focus on positivity and invert the contextual meaning of a particular problem into an opportunity to innovate positive change (Fredrickson & Losada, 2005; Roepke & Seligman, 2015).

The second property of macromanagement is prospection. Prospection is the capability of being future-focused driven by optimism. Combined with a positive outlook, prospection serves as a mechanism for leaders, managers, and other stakeholders to think by optimism and forwardly productive management; and this way of thinking and doing exerts an optimistic future-focused controllability within social environments (Na et al., 2019). In contrast, backward thinking and management with pessimistic outlooks of social environments have languished stakeholders' controllability and capabilities associated with negative effects and senses of hopelessness and helplessness in social environments (Maier & Seligman, 2016; Seligman et al., 2013). This insight into prospection has been studied in social environments, such as city ecosystems, and has been found to be an instrumental social managerial capability when assessing and implementing public policy for socio-economic well-being indicators (Adler & Seligman, 2016; Roepke & Seligman, 2015). The macromanagement meta-epistemological elements of wholeness (vertical and horizontal polycentric action) with positive future-focused approach and outlook is the essential management strategy for multi-stakeholder processes and integrative ecosystem management for the developmental goal of CASE.

This third management organizing form has emerged with a significant number of leaders and stakeholders that think small reconfigurations have the potential to produce large systematic changes (Cooperrider & McQuaid, 2012; Senge, 1990) ; and where a small shift or innovation in one thing can produce big changes in everything (Meadows, 1999). Meadows' (1999) small shift approach to systemic change relies on managerial hierarchic organizing form and stakeholders' local competencies and capabilities to recognize major leverage points of the system and move that leverage point in the right direction. Innovation is widely considered to hold the keys for whole systems sustainability, but particularly in cities, communities, and ecosystems (Berkowitz &

130

Dumez, 2016; Boons & Lüdeke-Freund, 2013; Cassiolato et al., 2014; Henrekson, 2014; Macharis & Kin, 2017; Schaefer et al., 2015). Some empirical research suggests that stakeholder engagement for systems-level sustainable innovation is achievable through its core leaderships' understanding of the cognitive, emotional, and behavioral dimensions of the stakeholder engagement (Jonas, Boha, Sörhammar, & Moeslein, 2018). Moreover, other scholars complement this notion of inclusive innovation, articulating that designing the appropriate configuration and aligning capabilities of stakeholders for city ecosystem and systems development is essential to achieving sustainable innovations (Berkowitz et al., 2017; Gulati et al., 2012), understanding how stakeholders in open-systems practices are empowered towards organizational effectiveness (Kirkman et al., 2011), and by following the generative guiding principles and processes of organizing for sustainability-driven innovation (Parrish, 2010). The macro-management objective is not intended to diminish key top management and decision-maker stakeholders' nor an organization's power; rather, the intention is for key management leaders and participating stakeholders to create network environments and platforms that promote inclusiveness by unlocking the greatest collective humanistic potential of ingenuity; which complexity theory informs us, "that it can emerge from anyone anywhere" (Mitchell, 2009).

Underlying this systems-level organizational design task is the adaptive learning processes involving skillful but imperfect rational agents and organizations (Holm, Lorenz, Lundvall, & Valeyre, 2010) and the diversity of stakeholders with informal and indigenous knowledge. Under this assumption, the diversity of organizations, institutions, and stakeholders within the city have untapped capabilities to enhance their individual competencies through searching and learning; and the systematic capacity building occurs in interactions between systems' stakeholders. What emerges because of this self-organizing systems approach develops into organizational gestalts of flourishing within the CASE. Gestalt in ecosystem designing is the organizing capability to approach a design task or inquiry creatively, individually, and collectively, yet maintain unity across the varying design outcomes (Yoo, Boland, & Lyytinen, 2006). This simultaneous co-existence of unity and variety reflects benefits for the whole outcomes in the form of innovation and new competencies recognized by stakeholders, organizations, institutions, and other subsystems from the innovative design and learning processes. The design gestalt is understood, as it relates to ecosystems development, as a "virtual" capability that combines ideas, values, resources, tools, and stakeholders into ensembles that can create and project remarkable multi-beneficial artifacts (Yoo et al., 2006). By form, characteristics, or a combination of both, these artifacts are produced as a representation of the organizing structures that designed them.

Resilience and the supporting theory emerged from scholarly discourse on general systems theory and have been studied in disciplines such as ecology, psychology, engineering, and sociology (Boschma, 2015). Resilience is defined as the capacity of a system to undergo change and still retain its basic function and structure (Folke, Carpenter, Walker, Scheffer, Chapin, & Rockström, 2010). As it relates to sustainability, resilience can be defined as the ability of organizations, communities, and systems to cope with external stresses and disturbances from social, political, economic, institutional, and environmental change. Evolutionary systems resilience, in particular, suggests that all systems are in a constant state of adaptation within an ever-flowing field of change (Davoudi, Shaw,

Haider, Quinlan, Peterson, Wilkinson et al., 2012; Simmie & Martin, 2010), and in the context of ecosystems resilience is constructed from the panarchic systems' dynamic capabilities of risk anticipation, reflexivity (adaptive learning), and responsiveness to perturbation. It is dependent on the resource accessibility; the diversity of stakeholders, leadership, education, and economies; the trust, cooperation, and coordination within stakeholders' engagements; and how stakeholders' decision-making adapts to perturbation, adversity, uncertainty, and conflict.

Empirical studies in management have proved several key factors that build resiliency, and a few are stakeholder relationships of openness, transparency, and trust; connectivity of people and resources; with continuity and stability within the dynamics of the city ecosystem that are reactive (flexible) and anticipatory (risk planned) (Bec et al., 2019; Dale et al., 2010). A diversity of capacity building, synthesizing stakeholder capabilities, and adaptive capacities to perturbations—or the emergences of deviations from system norms— can influence the robustness of resilience building (Dale et al., 2010). Bec et al. (2019) added to Dale expressing that resource accessibility, diversifications of stakeholders, adaptive capacities capabilities in continuity, the right "mix" of high-quality leadership (clout), informal leadership that builds (trust), and engaging peripheral stakeholders that have competing institutional logics can connect and build systems resiliency through cooperation.

Although scholars have developed the key factors leading towards sustainable innovation and resiliency, a social structuring and integrative process that describes a "how to" achieve a flourishing ecosystem on the scale of a city is less developed. The functionality of AI-SOAR offers the operationalized integrative systems management and social structuring that yields positive impacts from co-collaborative efforts toward the creativity and resiliency desired to be achieved. AI-SOAR is macro-managing complex systems, or in ecosystems language, a living system affecting a higher-order living system. It is important to note that city-level sustainable development achievement should not be viewed as a linear pathway to a sustainability goal, but its non-linear nature requires the integral tasks to be viewed as a change-stability balance-seeking temporal process where a mass humanistic flourishing state can emerge (Kemp & Loorbach, 2003). The developmental state of flourishing in city ecosystems in both social and physical systems can imply how well developed the evolutionary fitness of the city ecosystem has progressed from a "survival" vitality to a "thriving" vitality.

***Operating the system and process of appreciative inquiry and the SOAR Framework.*** The application of AI summits and platforms has emphasized the concept of positive potential—the best or what has been, what is, and what might be. AI is a multi-stakeholder process of positive change (Whitney & Trosten-Bloom, 2010) that reframes traditional managerial problems with a salutogenic approach to team, group, organizational, and systematic vitality. Centered around five key principles, the philosophy of AI is grounded on the macro-management's ability to create change with generative inquiry, amplify organizational strengths, harness the transformative power of the 'positivity ratio' with small positive systematic shifts, imagine in wholeness "what is best," and develop a wider lens of what is to be appreciated or valued (Cooperrider & McQuaid, 2012).

The challenges of operationalizing AI may, at first, seem too complex to manage for achieving CASE innovative performance and resiliency; contrarily, it does not. Some of the problems require only small systemic innovations or levers, which enable the fulfillment of needs in an entirely new manner. Yet, planning is difficult when things useful to us today may be of no use in the future, and things we do not value may be essential to humans living in the future (Gowdy, 1994). AI platforms are instrumental in creating a system where innovation can emerge from everywhere as we look to increase sustainable value in CASE. Stakeholders within CASE absorb creative potential energies by developing design-inspired collaborative multi-stakeholder tasks. Although it is not necessary for every single stakeholder in the city or ecosystem to participate; it is highly encouraged to integrate the collaborative capacity of multi-disciplinary groups in practice and interest, as well as representation from diverse sets of city ecosystem settings within a CASE, including situations of competing institutional logics. Strategic innovation is the result of bringing a diverse set of voices into the strategy dialogue, among other issues (Hamel, 1998). Dalton and Dalton (2006) echo this finding that minority opinions stimulate creativity and divergent thought, which, through their participation, can manifest as innovation. The utilization of big data, social media, and online groups have all contributed to nascent emergences through social structuring and positive action. The digitalization of data mining, social organizing systems by social movements has facilitated the extension of stakeholder engagement aiding in the discovery of innovative solutions beneficial to communities and cities through the network of peripheral stakeholders from other organizations, communities, systems, or other CASE (Burt, 1992; Steele, 2017).

As AI is applied, macro-managers utilize the 4-D cycle with generative questions designed to activate the interdependent design task of stakeholders. These macro-managers start by pre-framing a powerful interdependent task with a purpose much larger than the system. Applying the term task assumes actions to be taken, dissimilar to a mission statement that often represents a static state of intent in collaboration. The key question in pre-framing the interdependent task is, "what do we want to create, not what do we want to solve or avoid?" (Cooperrider et al., 2012). This interdependent task in CASE flourishing context can be an important opportunity that designs the task from those key questions and has potential implication on all diverse sets of stakeholders within the system. After the powerful task is established by key asset orchestrators, the first phase of the 4-D cycle is the "Discovery" (tasked with appreciating in value "what is") next is the "Dream" phase (imaginative or envisioning the future stage), followed by the "Design" phase (Co-creating the dream), and the "Destiny" phase (empowering, adaptive learning, and sustaining stage) (Cooperrider et al., 2003). AI invites people to engage in a shared dialogue about positive images and affirmations to create new and alternative possibilities in their communities and cities. Macro-managers facilitate shared dialogue among stakeholders in an organization or city ecosystem that focuses on what works well rather than what is not working well or details to deficiencies (Whitney & Fredrickson, 2015)). The shared dialogue occurs through AI-based conversations about gaps, possibilities, and a desired future (Bushe, 2012; Stavros et al., 2018).

Macro-management applied as AI brings out the best in human social systems, reaching across resource-rich boundaries sharing leadership and ownership of an interdependent task giving a sense of empowerment, and resulting in more effective collaborative efforts to appreciating—in a duel meaning of valuing and accumulating value—creativity and resilience in CASE. Unlike traditional system-wide transformative efforts that execute strategies within isolated silos, departments, or organizations, AI operationalizes collective stakeholders' brainpower by creation tasks from modularity, configurations, combinations, and interfaces with systematic acceleration and scaling up solutions (Cooperrider & Zhexembayeva, 2012). The concentration of effects that emerge from AI participation spreads throughout the system causing dissipative effects of systematic resiliency and innovation. This effect is also aligned with Holling (2001) and Simon (2019) explanation of an adaptive system as it relates to the significance of hierarchal structures, their effects on the higher-ordered structures, and the dynamic functionality that shapes both sub-system and ecosystem.

The strengths, opportunity, aspiration, results (SOAR) framework has its origins within the AI practitioner and academic scholarship; and the framework shares AI's five principles with underlying values of positive psychology, dialogic communication, generative conversations, and generative results (Bushe, 2012; Bushe & Paranjpey, 2015; Maxton & Bushe, 2018; McClellan, 2007; Stavros et al., 2018; Whitney & Fredrickson, 2015). The SOAR profile measures both AI and the four elements of SOAR (strengths, opportunities, aspiration, and results). The framework integrates positive psychology and AI through its focus on positive deviance (i.e., moving toward positive energy and away from negative energy), its encouragement of possibility thinking through generative conversations, and the formulation and implementation of a positive strategy by identifying strengths, building creativity in the form of opportunities, encouraging various stakeholders to share aspirations, and determining measurable and meaningful results (Cameron, Dutton, & Quinn, 2003; Cameron & Spreitzer, 2011; Stavros, Cooperrider, & Kelley, 2007). It is assumed multi-stakeholders are part of the collaborative effort, and all stakeholders are represented as the ecosystems' integrative management strategy. SOAR builds positive psychological capacity by using dialogue and generative conversations to transform strategic thinking, planning, and leading (Cole & Stavros, 2019). Through its focus on positive deviance, SOAR integrates the philosophies of AI and positive psychology by framing strategy as a set of processes that enable collective resourcefulness and generative dynamics that lead to positive states or outcomes (Fredrickson, Cohn, Coffey, Pek, & Finkel, 2008). Starting the SOAR process first focuses on finding and inquiring *strengths* or assets already in possession of an ecosystem. These assets include natural, financial, human, knowledge, tangible, intangible, and other resources available to the ecosystem. The next step sets the precedents of imaging opportunities and visualizing collectively desired outcomes. The *opportunities* step is a critical deviance from traditional intervening strategies that problematize. Instead of a perspective of negativity, weaknesses, deficiencies, and threats: stakeholders can flip problems into opportunities. For example, instead of framing the wicked problems of joblessness and affordable housing, participating stakeholders can generate dialogue around the opportunity and reorganize their thoughts and assets towards envisioning new job possibilities and co-creating the values of accessing housing based on essential and desired needs of stakeholders. This leads to the next step of *aspiration* that establishes the tactical functional plans and couple them with strategic initiatives to integrate programs that function to reach the planned future outcome. And lastly, the *results* step is the actional step that inspires stakeholders to implement and achieve the desired outcomes. An illustration of this SOAR framework process flow is shown in Figure 8.

Figure 8: The SOAR Framework[2]

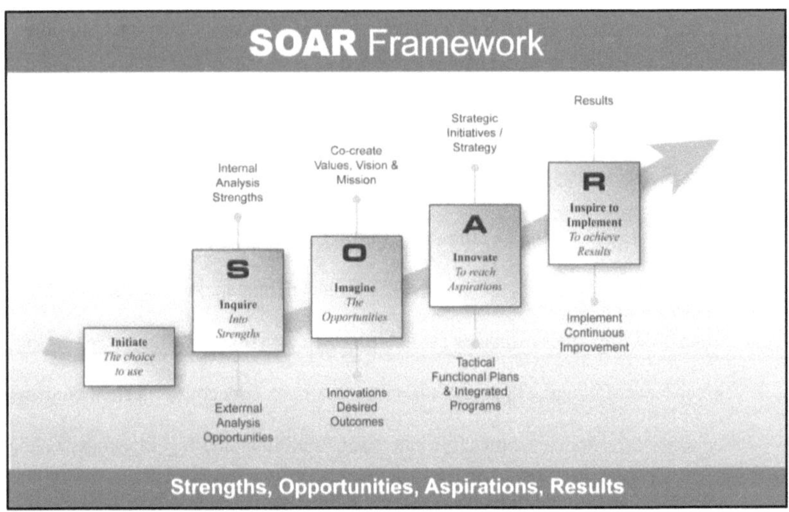


In the context of CASE's collaborative capacity building, AI and SOAR can be best described as the combination of organizational structure and organizational structure functional process (Herrmann, Jahnke, & Loser, 2004). Starting the operation of the AI platform, CASE's hierarchal organizing form structures architects, developers, and managers as holding power, leadership, and relative resources. The collaboration and processes within the SOAR process are intended to be inclusive by the representation of diverse social classes, social cognitions, genders, business scales, institutions, cultures, and ethnic backgrounds within the ecosystem. It is also critical for the organizational structure to have participating stakeholders from the Base of the Pyramid (Cañeque & Hart, 2017) to asset orchestrating management scattered towards the top. It is the presence of diversity as well as the collaborative effort, regardless of the interdependent task created by orchestrating management; if the positive actionable task at its core is planned accordingly in the design method of AI, the execution of the process during an AI Summit should yield positive emergences of innovative ideas or capacities that can benefit the CASE co-creating a thriving city ecosystem vitality. Therefore, my assumption that the operating system and process of AI-SOAR leads to an increase of innovative capacities amongst participating stakeholders, which in turn, also stimulates the robustness for resilience within the ecosystem to support a flourishing city ecosystem vitality. Thus, I hypothesize:

> *Hypothesis 1. Applying the AI platform with the SOAR framework approach increases innovative capacity.*
>
> *Hypothesis 2. Applying the AI platform with the SOAR framework approach increases resiliency.*

Tethering from Dale et al. (2010), I echo the same assumption in reference to flourishing city ecosystem (CASE) vitality and underpinning relationship between innovation and resilience. They claim there are dynamic capabilities occurring between the two. One dynamism through building resilience and the other dynamism of building innovative capacities; and when combined emerges a flourishing ecosystem. I understood the use of the word "dynamism" had meaning related to the energy, action, process, and forces that move by path or trajectory; usually, in a nonlinear way. Interpreting dynamism within this framing, I understand this dual effect as causal forces within the whole systems, where the sub-ecosystems of innovation have a positive influence on the sub-ecosystems of resiliency, and vice versa, in a symbiotic relationship. The macro-management system and process AI-SOAR has the potential to facilitate the transformation within the whole system and dissipate throughout the ecosystem. The AI-SOAR macro-management system relies on humanistic capabilities of collective planning, designing, implementing, and evaluating by designing protocols that promote a social constructing—with a positive bias—to affect all panarchic subsystems, including social systems. Through positive framing, open communication, and focusing on stakeholder needs and innovative solutions that are both generative and build meaningful collaborative relationships, I pose the AI-SOAR management system as the driver that enables the capacity growth for innovation and a robust response of resilience. Since I propose the combination of AI-SOAR is an antecedent to increasing the city ecosystem vitality, I also posit that there is a double mediating effect of building resiliency and innovative capacity. Moreover, I theorize that one has a positive influence on the other, resulting in the dissipative emergence of flourishing CASE vitality (Dale, Foon, Herbert, & Newell, 2015). Therefore, I hypothesize:

> *Hypothesis 3. Innovation capacity increases the effectiveness between applying the AI platform with the SOAR framework approach and resiliency towards a robust city ecosystem vitality.*

> *Hypothesis 4. Building resilience increases the robustness between applying the AI platform with the SOAR framework approach and innovation capacity towards a flourishing city ecosystem vitality.*

Figure 9 presents a visual summary of the Appreciative Inquiry and SOAR platform's influence on the city ecosystem's vitality in my hypothesized model.

Figure 9: Hypothesized Model

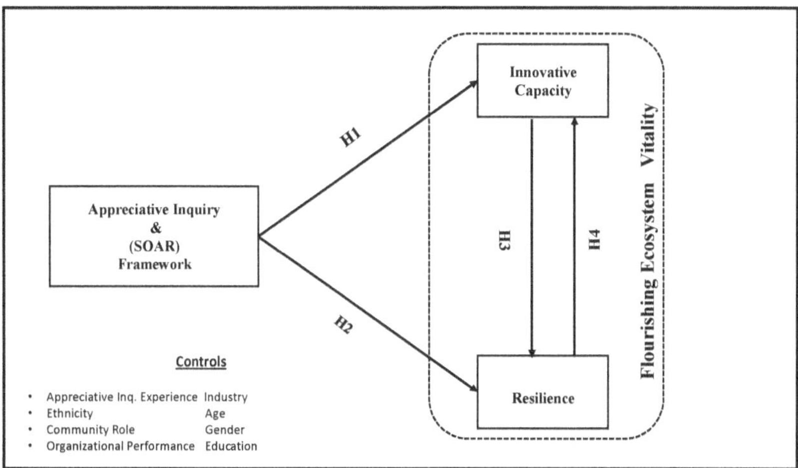


## Research Design

### *Measures.*

**Resilience (RES).** Resiliency was measured with an eighteen-item scale and asked during the survey. The 18-item scale was adopted from Bec et al. (2019), and these performance indicators of resilience formed from the four-dimensional concepts of *Resource Accessibility*; *Diversity*; *Clout, Trust, and Cooperation*; *and Capabilities, Coordination, and Continuity*. CASE resilience can be defined as a positive psychological system capacity to rebound, to "bounce back" from adversity, uncertainty, conflict, failure, or other perturbations. The survey items were measured on a five-point Likert scale, and subjects were asked, "On a scale of 1–5, with 5 being the highest score: compared to other cities, how would you perceive your city's successful pursuit of the following performance indicators?" Each of the sub-scale items of resilience was as follows: (1) *Resource Accessibility* (a) Can access funds for dealing with short-term disasters, (b) Can access insurance coverage for major public and private assets; (2) *Diversity* (a) Has a diverse economy and workforce, (b) Has opportunities for education, training, and learning; (3) *Clout, Trust, and Cooperation* (a) Has leaders who adjust quickly to change, (b) Has strong leaders who work well together, (c) Is made up of people who support each other, (d) Is made up of people who trust each other; (4) *Capabilities, Coordination, and Continuity* (a) Has long-term plans aimed at ensuring a diversified local economy, (b) Has long-term plans that aim to manage natural resources sector development, (c) Has long-term plans that aim to manage sustainable development, (d) Integrates and shares knowledge amongst stakeholders, (e) Is regularly informed about changes affecting the community, (f) Participates in risk and vulnerability planning, (g) Plans for disasters, loss, hazards, vulnerabilities, and risk, (h) Prepares and trains for long-term change, (i) Prepares and trains for short-term change, (j) Works well together across internal and external bodies. The coefficient alpha reliability estimate of this scale was 0.84.

**Innovative capacity IC.** Innovative capacity, or creative capacity, was measured with a 14-item scale adopted from George and Zhou (2001) and Scott and Bruce (1994a). Innovative capacity, also understood as creative performance, was defined as the managing of creativity and innovation outcomes through stakeholders' cognitive energy, reward, recognition, and objective satisfaction. Participants were asked on a five-point Likert scale, "On a scale of 1–5, with 5 being the highest score: Compared to other cities, how would you perceive you and other stakeholders in your city's successful pursuit of the following performance indicators?" The 14 items were as follows: "Stakeholders in or geographic community…." (a) suggests new ways to achieve goals or objectives, (b) come up with new and practical ideas to improve performance, (c) are good sources of creative ideas, (d) are not afraid to take risks, (e) search outside our network for new technologies, processes, techniques, and/or product ideas, (f) suggests new ways to increase quality of life, (g) often have fresh approaches to issues, (h) develop adequate plans and schedules for the implementation of new ideas, (i) promotes and champions ideas to others (j) often have new and innovative ideas, (k) come up with creative solutions, (l) suggest new ways of performing task, (m) exhibit creativity when given the opportunity, (n) are good sources of creative ideas. Of the fourteen items in the survey, two items were adopted from Scott and Bruce (1994a). The coefficient alpha reliability estimate of this scale was 0.81.

**Appreciative inquiry AI.** Appreciative Inquiry was measured with a six-item scale adopted from Cole and Stavros (2019). AI is defined as a whole systems strategy model to strategic thinking, planning, and leading to positive change for the whole. The AI model focuses on strengths and opportunities instead of weaknesses and problematic issues for system-level solutions. Participants were asked to select one of the response options from "never" to "always" ranging "Never," "20% of the time," "30% of the time," "40% of the time," "50% of the time," "60% of the time," "70% of the time," "80% of the time," "90% of the time," and "Always." The six items representing AI were as follows: "When you approach strategy in your life, team, organization, or community, how often do you focus on..." (a) Collaborative Relationships, (b) Generative Questions, (c) Open Communications, (d) Positive Framing, (e) Stakeholder Needs, and (f) Whole systems. Each of these represents theoretical characteristics of AI, and a definition was provided of each characteristic for the participant's comprehension of the concept. The

coefficient alpha reliability estimate of this scale was 0.86.

**Strengths, opportunities, aspirations, results framework SOAR.** The SOAR

framework was measured with a 13-item sub-scale formed from four-dimensional

concepts: (1) Strengths, (2) Opportunities, (3) Aspirations, and (4) Results. The scale was

adopted from Cole et al. (2019). SOAR is defined as a strengths-based framework with a

participatory approach to strategic analysis, strategy development, and organizational

change. Participants were asked to select one of the response options from "never" to

"always" ranging "Never," "20% of the time," "30% of the time," "40% of the time,"

"50% of the time," "60% of the time," "70% of the time," "80% of the time," "90% of

the time," and "Always." The four sub-scale items representing SOAR are as follows:

"When you approach strategy in your life, team, organization, or community, how often

do you focus on..." (1) *Strengths:* (a) Assets, (b) Capabilities, and (c) Strengths; (2*)*

*Opportunities*: (a) Ideas, (b) Opportunities, (c) Possibilities, and (d) Solutions; (3)

*Aspirations*: (a) Aspirations, (b) Desires, and (c) Values; and (4) *Results*: (a) Outcomes,

(b) Results, and (c) Completed Tasks. The coefficient alpha reliability estimate of this

scale was 0.83. The table below displays the survey instrument developed with the

number of indicators and citations, and a copy of the construct table can be found in

Appendix F.

**Table 6: Survey Indicators**

| SOAR Towards Sustainable Innovation and Resilience Survey Indicators | | | |
| --- | --- | --- | --- |
| **Variable** | **Items** | **Type** | **Citation** |
| Strength, Opportunities, Aspirations, & Results Framework SOAR | 13 | Reflective | (Cole, Cox, & Stavros, 2018; Cole & Stavros, 2019) |
| Appreciative Inquiry AI | 6 | Reflective | (Cole, Stavros, & Zerilli, 2017) |
| Resilience (RES) | 18 | Reflective | (Bec et al., 2019) |
| Creativity (CR) | 14 | Reflective | (George & Zhou, 2001; Scott & Bruce, 1994a) |

***Sample.*** The data were obtained as part of a new study on ecosystems-level strategies towards flourishing, which draw upon various stakeholders in cities and communities for increased degrees of self-organizing, inclusion, innovation, and capacity-building efforts. During the study, I measured individual's responses of AI and SOAR as reported on "other participants" in the community. I measured participants' ratings of creativity and the resilience of others within the community. The survey instrument was slightly modified from the AI and SOAR survey instrument of Cole et al. (2019) to reflect items describing another focal person's behavior rather than their own behavior. Respondents rating someone else, such as their leader, or someone else whom the respondent is familiar with, may perceive traits and competencies in others better than ratings the individual makes in him or herself (Libbrecht et al., 2010; Van der Zee et al., 2002).

I developed my demographic requirements to be inclusive of various types of categorical stakeholders that are relevant to the boundaries of a city and range from top-level formal management leadership to active volunteers. Survey items were also added to collect and identify the diversity of stakeholders by formal and informal arrangement, geographic location, age, and gender, as well as level of station and community outreach, specific tasks assigned, and education. Detailed illustrations of individual demographic characteristics can be found in Appendix G.

*Procedures.* The subject sample population was collected and developed through my past experiences in several professional organizations and the professional network as a business consultant in sustainability. Three waves of email distributions were performed, with a total of 500 emails disseminated to prospective professionals within the network. I also disseminated the survey through social media outlets such as Facebook and LinkedIn to reach the participants in my network. This method invited over 10,000 samples within the professional network to participate in the survey. Consent forms were embedded within the Qualtrics platform and collected before the online survey was initiated. A total of 44 samples were collected during this phase of sample collection. Subsequentially, the challenges of voluntary survey participation through my network were found inadequate in accumulating enough samples to reach a significant sample effect. Therefore, I utilized MTurk to drive the sample count and reach a significant sample effect necessary to test the model. And as a result, I collected 44 panel samples from my professional network and purchased 300 panel data samples for a total of 344 samples within the targeted demographics. Of the 344 individuals who opted to participate, three subsequently were missing data, incomplete, or unusable; hence, a total of 341 participants responded for a 3% aggregated response rate.

The sample respondents were 55.4% male, 43.7 % female, 0.6% non-binary, and age 90.3% was predominately between the range of 23–57 years old. The greatest age range responded 29.9% were found in the 28–32 age group. Around 90% of these participants answered as an employee within some form of industry versus 10% that responded as volunteer or student. Over half (57.2%) of the respondents earned a bachelor's degree, 22.3% carried a master's degree, 7.6% a high school diploma, 6.5% an associate degree, and 6.5% a doctoral/professional degree. As it relates to community role, 62.2% responded as an active member of their community in some capacity, 10% responded as a community leader, and 27.9% answered as not actively working within their community. Table 7 summarizes the demographic characteristics of the study sample.

**Table 7: Demographic Characteristics of Study Sample**

Demographic Characteristics of Study Sample

| Characteristic | n | % | Characteristic | n | % |
|---|---|---|---|---|---|
| **Total** | 341 | 100 | **Ethnicity** | | |
| **Age** | | | *White* | 182 | 53.4 |
| *18-32* | 167 | 49 | *Non-White* | 159 | 46.6 |
| *33-57* | 147 | 42.8 | | | |
| *Greater than 57* | 27 | 7.9 | **Community Role** | | |
| **Gender** | | | *Non-active community member* | 95 | 27.9 |
| *Male* | 189 | 55.4 | *Active community member* | 212 | 62.2 |
| *Female* | 149 | 43.7 | *Community Leader* | 34 | 10 |
| *Other* | 3 | 0.9 | | | |
| | | | **Professional Domain** | | |
| **Education** | | | *Organization position* | 250 | 73.3 |
| *High School* | 26 | 7.6 | *Government position* | 43 | 12.6 |
| *Assoc. Degree* | 22 | 6.5 | *Other* | 48 | 14.1 |
| *Bach. Degree* | 195 | 57.2 | | | |
| *Master Degree* | 76 | 22.3 | **Organizing Performance** | | |
| *Doc. or Prof. Degree* | 22 | 6.5 | *Performance declined* | 37 | 10.9 |
| **Industry** | | | *Performance not changed* | 110 | 32.3 |
| *Industrial* | 307 | 90 | *Performance improved* | 134 | 39.3 |
| *Other* | 34 | 10 | *Not applicable* | 60 | 17.6 |

For this study, I included the controls of Appreciative Inquiry experience (AIEXP), Ethnicity (ETHNIC), perceived organizational performance (ORGPRFM), and city ecosystem role (COMMROL) into this model. I controlled for AIEXP to observe if there was a significant difference in responses between participants that have experienced or have a competency of Appreciative Inquiry. ETHNIC and COMMROL were included to observe a significant difference amongst various ethnic groups and active roles in their city ecosystem. Lastly, ORGPRFM was integrated into the model to observe if there were significant differences in responses due to participants' perceptions of their city ecosystem's performance.

*Analyses.* Using SPSS, I analyzed my data to produce data frequencies and descriptive statistics, then computed a syntax in Mplus to perform a confirmatory factor analysis (CFA). The CFA was performed as an assessment and output of confirming the measurement model, determining model fit, convergent and discriminant validity, composite reliability, factor correlation matrix, structural equation modeling SEM, and testing mediating effects. I utilized SPSS to conduct EFA. First, I analyzed the sample data by descriptive statistics and dimension reduction computation. I set the factor statistics to report KMO and Bartlett's test of sphericity as well as a reproduced correlation matrix. Method of extraction was set to Maximum Likelihood (ML) to analyze the correlation matrix, then setting the extraction based on eigenvalues greater than one. I also set the extraction rotation to Promax and suppressed small coefficients to 0.3 to examine the factor structure (Costello & Osborne, 2005) and executed the analysis. Subsequently, I adjusted the Kaiser Normalization value several iterations to perform the extractions with a fixed number of factors at three, four, five, six, and seven. This method was used to maximize differences between factors and provides a model fit in the estimate at a fixed factor of four (Hair, Black, Babin, Anderson, & Tatham, 1998) representing my four latent variables. After running these analyses, I reviewed the KMO and Bartlett's Test results, communalities, total variance explained, and best converging loadings pattern matrix, which can be found in Appendix J. The KMO and Bartlett statistical value was 0.97, which is greater than 0.70 and meets the threshold of acceptability (Matsunaga, 2010). After reviewing the communalities, I found no item loadings lower than 0.450 and 0.488, indicating that beyond these two measurement items, I have minimal concerns with any factor loadings < 0.500, which is the acceptable threshold. Although my intention was to expect four unobserved variables; AI, SOAR, RES, and IC, I could not achieve discriminant and convergent validity that resulted with the two unique unobserved variables of AI and SOAR separately. I went back to the study that the AI and SOAR scales were acquired (Cole & Stavros, 2019), and reviewed the data results of that study to determine where the separation between these two variables become problematic. I discovered when AI was tested as a positive predictor of SOAR, the effect was 1 (standardized b=1.001, p<0.01), indicating that these two variables may risk being too similar and converged as a second-order construct. This second-order construct was labeled Appreciative Inquiry and SOAR, or AI-SOAR (AIS). By merging AI-SOAR as one latent variable, along with innovative capacity and resilience, explains a cumulative variance of 54.3%, which is over the >50% acceptable threshold, see Appendix H. The pattern matrix results with three factors revealed convergent validity and reliability as most loadings appear to have values greater than 0.50, and ideally greater than 0.70. However, there were four item factors loadings on Resilience that were noticed to be <0.50, but these values were considered acceptable due to the sample size being >300 respondents (Hair et al., 1998). There were no significant cross-loading factors found in the pattern matrix, nor were there any cross-loadings with a value difference <0.20 across unobserved variables. The Cronbach's $\alpha$ for each unobserved variable was calculated, and all resulted in values greater than 0.7, AISOAR

0.962, RES 0.904, IC 0.795; thus, confirming reliability of this measurement model. Correlation results between the unobserved variables can be found in Appendix J. These unstandardized results reveal AISOAR- IC 0.680, AISOAR-RES 0.732, IC- RES 0.744. These factor correlation values also raise concerns, as all are >0.70, and may become problematic to establish both discriminant and convergent validity of the measurement model.

For the confirmatory factor analysis (CFA), I continued our assessment of convergent and discriminant validity, composite reliability, the factor correlation matrix, goodness-of-fit, SEM, and mediating effects in MPLUS. The CFA syntax in Mplus was computed to compare our results of the descriptive statistics and factor analysis and to confirm similarities between SPSS and Mplus results. The SPSS data was converted to Mplus translation with each measurement item transposed into a syntax recognizable by Mplus, then standardized and saved as a *.dat file. I re-analyzed our data in Mplus by running an EFA to test 3-7 factors. Extraction method was defaulted using maximum likelihood as the estimator along with a GEOMIN rotation. Result outputs were similar in value between SPSS and Mplus, and the three factors measurement model revealed the best fit, convergent validity, and reliability. The goodness-of-fit indices can be viewed in Table 8.

**Table 8: Model Fit Indices**

| | | | | |
|---|---|---|---|---|
| CMIN | - | 1912.9 | --- | --- |
| dF | - | 971 | --- | --- |
| CMIN/dF | <3 | 1.970 | GOOD | Byrne (2016); Hair, Black, Babin, and Anderson (2010) |
| p-Value | >0.05 | 0 | GOOD | Byrne (2016) |
| RMSEA | < .08 | 0.053 | GOOD | Hu and Bentler (1999) |
| SRMR | < .09 | 0.038 | GOOD | Hu and Bentler (1999) |
| TLI | > .90 | 0.914 | ACCEPTABLE | McDonald and Marsh (1990) |
| CFI | > .90 | 0.910 | ACCEPTABLE | Hu and Bentler (1999) |

All goodness of fit indices: $x^2$/df =2.039, TLI=0.896, CFI=0.903, RMSEA=0.055, SRMR=0.039 meets the acceptable requirements and indicates this measurement model has a good fit (Byrne, 2004; Hu & Bentler, 1999; Kline, 2011; McDonald & Marsh, 1990; Podsakoff, MacKenzie, & Podsakoff, 2012). The factor correlation was developed and integrated into a table with composite reliability (CR), average variance extracted (AVE), and Maximum Shared Variance (MSV). The CR and AVE were calculated in the CFA, and MSV was calculated in MS Excel from extracted CFA output results. Table 9 displays the results of the factor correlation, CR, AVE, and MSV.

**Table 9: Factor Correlation, CR, AVE, & MSV Results**

| Factor | CR | AVE | MSV | AI-SOAR | Innovative Capacity | Resilience |
|---|---|---|---|---|---|---|
| AI-SOAR | 0.962 | 0.567 | 0.663 | **0.753** | | |
| Innovative Capacity | 0.959 | 0.629 | 0.723 | 0.745** | **0.793** | |
| Resilience | 0.935 | 0.492 | 0.585 | 0.788** | 0.823** | **0.670** |

As displayed, CR values for AI-SOAR=0.962, IC=0.959, and RES=0.935 affirm reliability with all factors greater than 0.7. The reliability values from the EFA

152

(Cronbach's a) and CFA (CR) confirm reliability; moreover, many method scholars consider CR values closer to "true" reliability of measurement (Peterson & Kim, 2013) and consider it a stronger indicator for reliability. The AVE values for AI-SOAR=0.567 and IC=0.629 revealed convergent validity, having values greater than 0.500; but the initial RES=0.449 resulted as being problematic and under the <0.500 threshold for convergent validity. My options to mitigate the model results with the goal of achieving discriminant and convergent validity with all variables was to review the Modification Indices (MIs) and identify problematic observed variables and residuals. After reviewing MIs, I determined that resilience items RES1-RES4, and RES7 were consistently observed as problematic from my analyses. Although it is not recommended to covary error terms, or remove observed variables, I made the decision to remove these items to achieve discriminant and convergent validity with the three factors. The measurement was close to achieving discriminate validity with AVE values >0.500 for all factors: AI-SOAR=0.567 and IC=0.629, and RES=0.492 as displayed in Table 9. The RES result was considered close to the acceptable threshold (Hair et al., 1998), trending towards convergent validity. The MSV results for AI-SOAR=0.663, IC=0.723, and RES=0.585. This informed us of the maximum of the variance shared between that factor and the other factor (Hair et al., 1998). Moreover, it is recommended that AVE is greater than MSV for discriminant validity (Hair et al., 1998). All MSV values were greater than AVE, which indicated no discriminant validity between the variables. However, I made the decision of a good model fit over attaining discriminant validity based on the Malhotra, Mukhopadhyay, Liu, and Dash (2012) argument that AVE is often too strict and recommends that reliability metrics can be established through CR alone. I do advise

153

viewing further outcomes, results, and claims with caution (Hair, Risher, Sarstedt, &
Ringle, 2019). I was marginally close to achieving discriminant validity, though items
were removed from the survey to strengthen the goodness of fit, validity, and reliability
of the measurement model.

Direct effects were tested to determine the causal relationship between AI-SOAR
with IC and RES, respectively. The direct effects were hypothesized by H1 and H2.
Mediating effects were tested using the Preacher and Hayes approach (Muthén, 2011;
Preacher & Hayes, 2008). The mediation variables IC and RES in Hypotheses 3 and 4
were analyzed using 2,000 bias-corrected bootstrapping resamples in Mplus. I created the
syntax for my model to produce direct and specific indirect mediating effects, including
controls, along with the confidence intervals bias-corrected two-tailed with 95% level of
confidence. From the Mplus output, I reviewed the $\beta$-weights and significance levels of
the direct and specific indirect relationships between the variables in the causal model.

**Results**

The results of the hypotheses testing with standardized $\beta$ weights, upper and
lower bounds, and p-values can be found in Table 10; and respectively, includes a
squared multiple correlation table to review. Table 11 displays the results and
significance of the controls within my model.

## Table 10: Direct and Indirect Effects with Squared Multiple Correlation

| Direct Effect Hypotheses Summary Table | | | | |
|---|---|---|---|---|
| **Direct Effect Hypotheses** | **Standardized β-Weights** | **Lower Bound CI[a]** | **Upper Bound CI[a]** | **Hypothesis Supported?** |
| H1. *Applying the AI platform with the SOAR framework approach increases innovative capacity for community vitality.* | 0.353* | 0.282 | 0.421 | Supported |
| H2. *Applying the AI platform with the SOAR framework approach increases resiliency for community vitality.* | 0.672* | 0.549 | 0.777 | Supported |
| **Indirect Effect Hypotheses Summary Table** | | | | |
| **Mediating Effect Hypothesis** | **Standardized β-Weights** | **Lower Bound CI[a]** | **Upper Bound CI[a]** | **Hypothesis Supported?** |
| H3: *Innovation capacity increases the effectiveness between implementing the AI platform with the SOAR framework approach and resilience towards a robust community vitality.* | Direct w/med: **0.775*** <br> Direct w/o med: **0.353*** <br> Indirect: **0.103*** | Direct w/med: **0.634** <br> Direct w/o med: **0.282** <br> Indirect: **0.022** | Direct w/med: **0.817** <br> Direct w/o med: **0.421** <br> Indirect: **0.209** | Supported |
| H4: **Building resilience** *increases the robustness between implementing the AI platform with the SOAR framework approach and innovation capacity towards a flourishing community vitality.* | Direct w/med: **0.733*** <br> Direct w/o med: **0.672*** <br> Indirect: **0.381*** | Direct w/med: **0.654** <br> Direct w/o med: **0.549** <br> Indirect: **0.322** | Direct w/med: **0817** <br> Direct w/o med: **0.777** <br> Indirect: **0.447** | Supported |

| Endogenous Variable | $R^2$ |
|---|---|
| **INOVCAP** (MED) (DV) | 0.693 |
| **RESIL** (MED) (DV) | 0.694 |

*$p<0.05$*, $p<0.01$**, $p<0.001$****

*a- Bias corrected two-tailed with 95% level of confidence*

## Table 11: Results of Model Controls

| | Estimate | S.E. | Est./S.E. | P-Value | Lower Bound CI[a] | Upper Bound CI[a] |
|---|---|---|---|---|---|---|
| **Controls** | | | | | | |
| **$y$ = RESIL** | | | | | | |
| AIEXP | -0.046 | 0.037 | -1.239 | 0.215 | -0.104 | 0.014 |
| ETHNIC* | -0.077 | 0.033 | -2.313 | 0.021 | -0.127 | -0.018 |
| ORGPRFM | 0.031 | 0.041 | 0.760 | 0.447 | -0.037 | 0.097 |
| COMMROL | -0.012 | 0.038 | -0.324 | 0.746 | -0.074 | 0.049 |
| **$y$ = INOVCAP** | | | | | | |
| AIEXP | -0.015 | 0.036 | -0.420 | 0.674 | -0.071 | 0.048 |
| ETHNIC | -0.001 | 0.028 | -0.043 | 0.965 | -0.049 | 0.043 |
| ORGPRFM | 0.019 | 0.039 | 0.478 | 0.633 | 0.019 | 0.081 |
| COMMROL | 0.031 | 0.034 | 0.904 | 0.366 | 0.031 | 0.091 |

*$p<0.05$*, $p<0.01$**, $p<0.001$****

As hypothesized, systematically organizing and operating the AI platform with

the SOAR framework process reveals a direct positive relationship to innovative capacity

($\beta$ = -0.353*), and applying the AI platform with the SOAR framework approach increases resiliency for ($\beta$ = -0.672*). Additionally, innovation capacity revealed a partial mediating effect between implementing the AI platform with the SOAR framework approach and resilience towards a robust community vitality ($\beta$ = 0.103*), and building resiliency also revealed partial mediation between implementing the AI platform with the SOAR framework ($\beta$ = 0.381*). As a result, Hypotheses 1–4 showed strong support for all direct and indirect effects with a ($p > 0.050$) significance. All estimates in hypotheses were bias-corrected two-tailed with a 95% level of confidence, and as shown, the upper and lower confidence intervals did not cross zero, which reinforces my confidence of each estimate with a 95% level of confidence.

I also analyzed variable variance in the causal model, and the results from the squared multiple correlations are attached under effect results and labeled Endogenous Variables. The endogenous variables within the structural model revealed some variances. Around 69% of the variance of the mediating and dependent variable innovative capacity explained by the model, and about 69% of the variance of the mediating and dependent variable in resilience was explained by the model. Controls were analyzed to determine if there were any control variables that had significant effects on the model and to realize if any control estimates indicated significance at the 95% level of confidence. Appreciative inquiry experience ($\beta$ = -0.046*ns*, $\beta$ = -0.015 *ns*), organizational performance ($\beta$ = 0.031 *ns*, $\beta$ = 0.019 *ns*), and community role ($\beta$ = 0.012 *ns*, $\beta$ = 0.031 *ns*) displayed minimal effects on resilience and innovative capacity; none were significant at the 95% level. What was also apparent in the upper and lower bound results was that these three controls crossed the zero-threshold, confirming no

156

significance in effects. The control of ethnicity ($\beta$ = -0.077*), or ethnic background, revealed an unexpectedly slight negative effect on resilience with significance at the 95% level. The confidence boundaries of the estimate were also between a threshold not crossing zero. In contrast, ethnicity ($\beta$ = -0.001 *ns*) presented no significant effect on innovative capacity, and the confidence intervals crossed zero confirming no significant effects due to ethnicity. Since there was a minimal effect between ethnicity and resilience, the size of the effect did not cause any alarms to my model; however, I do plan to take a closer look at this sample demographic effect in future research by evaluating my model in a multi-group analysis. The results of the hypotheses and final model can be found below.

Figure 10: Final Model with Results

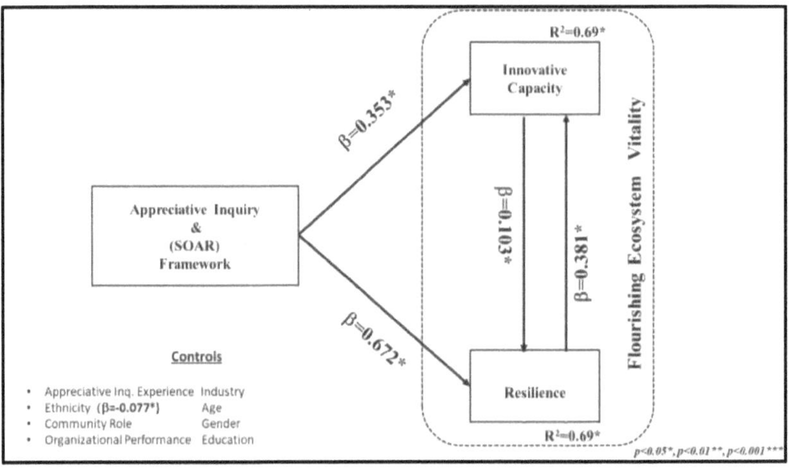

**Discussion**

This paper develops a structural model using several sets of indices to measure the macromanagement approach by a social-organizing structure and a whole systems strengths-based approach towards building innovative capacity and resiliency in CASE. From the perspective of individuals representing the diversity of individuals, organizations, and institutions, and meta-organization of these nested systems (Holling, 2001), I extracted the internalized perceptions from the human experience and uncovered how the underlying panarchic systems collectively within cities as ecosystem planned, strategized, and implemented themselves and others as an external materialization of their city ecosystem.

The SEM analysis confirmed that my structural model has reliability, discriminant validity, and partial convergent validity, and several indications that the causal logic model designed had a good fit. My model also confirms that AI-SOAR framework can be a useful tool under a macro-management organizational structure for transitioning the ecosystem, which, in addition, addresses underlying socioecological and socioeconomic concerns towards flourishing. The AI-SOAR framework creates the space and strategy for genuine collaborative dialogue that opens new design alternatives; while simultaneously self-generating the local social, environmental, and economic shape and form for present and future innovation on the scale of the ecosystem.

I synthesized a whole systems strengths-based integrative approach to collectively design and reconfigure a city as a sustainable ecosystem with building innovative capacity and resiliency index instruments. The scales and index used in the model captured seemingly complex and multi-dimensional dynamic capabilities with sub-dynamic capabilities that have strong relationships between AI-SOAR, innovation, and resiliency, which ultimately emerges as a gestalt of a flourishing community vitality. The key constructs of aspiration, stakeholder strengths focused, positivity framing, possibilities, generativity, performing interdependent task, opportunity, connectivity, collaboration, creativity, coordination, innovative ideas, inclusiveness, clout (power), diversification, trust, asset and resource allocation, planning, and solutions-based outcomes were embedded within my survey instrument, and represented the multiple dimensions and dynamisms related to my structural model, and the ecosystem. The key dimensions were aligned with the characteristics of a macro-management approach to co-designing by cultivating instrumental stakeholders' relationships and ideas, and externalizing the process to result in innovative artifacts that are intentionally purposed for present and future utility. Executing the process effectively, most of the artifacts should emerge as sustainable, generative, and beneficial to the ecosystem, as well as the diversity of stakeholders within, assuming there is inclusiveness or some form of representation of that stakeholder, organization, or participant in the overall process.

In this quantitative analysis, I reinforced the notion of Dale et al. (2010) and their observation that there are dynamisms between building resilience and building innovation capacities. I found that these two constructs are highly correlated but different enough to where one has a positive effect on the other; and from this dual effect develops the organizing gestalt of a flourishing ecosystem. This is an alternative that challenges the dominant classical form, structures, and gestalts of disproportioned modular areas of flourishing and non-flourishing scattered throughout the ecosystem.

The classical change management approaches to city development, environmental management, systems management, and social behavioral interventions have proven, at best, a 30% success rate of implementation. Management approaches for cities as ecosystems that focus on top-down controls of integrated value systems, and that only offer fragmented linear solutions based on individual, modulated, or nested problematic occurrences will continue to be stuck in a quandary of panarchic deterioration, resource depletion, and extinction. Moreover, problem shifting to future generations and short-term fragmented interventions that are usually incrementally effective and mostly ineffective are the opposite of a flourishing system. Thus, I engage scholars, architects, developers, and society, to employ an inclusive, generative, whole systems strengths-based approach—with a positive bias—to self-organize and meta-organize their cities as sustainable ecosystems. In this study, I offered and tested a conceptual model for stakeholders, specifically, leading stakeholders, to consider a perspective of understanding macromanagement for city ecosystem transition towards sustainability. In addition, this model can be an instrument for government officials, managers, leaders, and primary stakeholders in cities to operationalize the macromanaging approach of an AI-SOAR platform and measure this model and survey instrument periodically in their ecosystems to assess and monitor their collective citizen perspectives towards fostering innovation and developing their flourishing city.

Although I validated AI-SOAR, a meta-organizing macromanagement platform, has positive effects on city ecosystem innovation and resilience building, this conceptual model does not fully capture the mechanisms and dynamics of city ecosystem sustainable development and achievement. Furthermore, the Cleveland ecosystem was the only city in my Studies 1 and 2 samples that I confirmed had employed an AI platform for their city sustainable development achievement. Thus, I set out to find the combination of factors that can predict whether a city ecosystem has high sustainable development performance and the combination of patterns that predict city ecosystem entropy or low sustainable development performance. I were able to incorporate the meta-organization as an identified factor predicting high performance, as well as the factor asset orchestration dynamic capabilities. These factors, or causal conditions, represent two of three of my unit theories, and to develop more factors in Study 3, I draw from theory to discover other factors that predict high performance in city ecosystems.

**CHAPTER 7: STUDY 3 – *FS*QCA: CONFIGURAL IDENTIFICATION OF
FACTORS RELATED TO CITY SUSTAINABLE DEVELOPMENT
PERFORMANCE TOWARDS FLOURISHING CITIES AS SUSTAINABLE
ECOSYSTEMS**

**Introduction**

The UN SDG US City report reveals that there is significant work to do if

sustainable cities and the sustainable development goals (SDGs) are to be achieved in the

U.S. On average, the top 100 major cities scored total scores of 48.9% in the 2019 report

(Lynch, 2019). Moreover, the global challenge of the COVID-19 pandemic has impeded

global sustainable development progress in some areas but has exacerbated other

sustainability challenges. This has led some scientists to argue that the pandemic has

pushed most 2030 goals back an estimated 18 years, just to get back on track to pre-

COVID progress (Sachs et al., 2020). Over the last three years[3] of city-level sustainable

development data capturing, researchers have found that few U.S. cities have shown

modest improvements in their total SDG achievement based on the cumulative

sustainable city score, while other cities have experienced little improvement. There are

some cities that have shown a decline in total score, which suggests sustainable

innovation and development failures, but more importantly, has steered towards the

deterioration of social well-being, vital economies, and healthy city ecosystems. Despite

the importance and need for cities to develop and diffuse sustainable innovation to

improve SDG total scores and promote sustainable development on the scale of a city

ecosystem, over half of U.S. major cities continue to fail at improving their total scores

and advancing their community development initiatives. As a yearly assessment guide,

---

[3] Years: 2017–2019. Most of my data was collected and evaluated prior to the COVID-19 pandemic.

significant improvements in SDG relates to high sustainable development performance in city ecosystems, while small, incremental to no score improvements relates to low-to-no sustainable development performance. The human transformation needed towards sustainable city ecosystems will be essential to sustaining a connected and flourishing community vitality (Dale et al., 2010). Therefore, there is a need to comprehensively and systematically understand how some city ecosystems are not improving performance scores and not avoiding negative outcomes, while other city ecosystems achieve significant improvements in sustainable development and innovation within their total SDGs performance scores. The latter is the transformation needed for cities to reach a flourishing developmental form and a flourishing city as an ecosystem (Edwards, 2009; Newman & Jennings, 2012).

Previous research has identified many factors that influence sustainable innovation and city ecosystem success towards sustainability. For example, Heaton and Teece (2019) point out the role of universities in leading the local innovation ecosystems' flourishing, which contributes to the city's promotion and development of sustainable products, processes, and services. The sustainable innovation created by stakeholders in innovation ecosystems can also lead to more sustainable development in infrastructure and large-scale social innovation projects that generate benefits for global societies. Other scholars have found that attributes of a meta-organization governance structure can be an effective instrument for promoting sustainable development performance on the scale of an industry or city ecosystem by merging organizational capabilities amongst multiple stakeholders (Berkowitz, 2018; Valente & Oliver, 2018). Progressive sustainable development performance in cities has also been linked to building capacities such as managerial and financial capacities (Saha, 2009; Wang et al., 2012); having a

comprehensive climate adaption plan (Cooper, Evans, & Boyko, 2009; Saha, 2009;

Yigitcanlar, Dur, & Dizdaroglu, 2015; Yigitcanlar & Kamruzzaman, 2015); and powerful

leadership (Bec et al., 2019; Heaton et al., 2019; Mousavi, Bossink, & van Vliet, 2018;

Rant, 2020; Saha, 2009; Schoemaker et al., 2018; Teece, 2007) taking the lead in

orchestrating assets and resources that encourages citizen participation, pools resources,

enables community building, as well as facilitating the shared vision of achieving

sustainable communities. Moreover, some research suggests that the internal social

mechanisms of a city, such as local citizens' attitudes towards sustainability, knowledge

and cultural experiences, and climate awareness plays a strong role in advancing

sustainable development performance and achieving sustainable cities (Bain et al., 2019;

Carley & Smith, 2013; Duxbury, Cullen, & Pascual, 2012; Portney, 2005; Prugh,

Costanza, & Daly, 2000; Rousseau et al., 2019; Sachs et al., 2019; Wang et al., 2012;

Watene & Yap, 2015). Influencing the political dimensions of sustainable development,

multi-level government support and intervention has been presented as being the policy

enforcing factor that stimulates sustainable development performance as well (Dunphy,

2000; Florini & Pauli, 2018; Lee & Petts, 2013; Prugh et al., 2000; Saha, 2009; Voegtlin

& Scherer, 2017).

Even though we have this body of knowledge, studies have found contrary results

for the same contributing factors. From her research on city networks, urban institutional

failures, and multi-level government relationships, Frantzeskaki (2019) suggests that

multi-level government support and policy interventions between local, state, and

national levels can inhibit and act as a barrier to local city network's efforts to promote

164

urban institutions like sustainability and cause institutional failures in city networks'
resilience building and performance. For example, city networks can "…create conditions
for urban institutional failures in cities due to temporality issues between the emerging
institutions and localized agency adoption, political tensions between local and
state/national levels that leaves cities to fill institutional voids where state and national in
needed, and also due to national or state level governments' eclipsing of local
governments' institutional change efforts by continuing to establish the antiquated and
obsolete institutions in place" (Frantzeskaki, 2019: 713).

Overall, analyzing factors in isolation has been insufficient to characterize the
complex causes of meta-organizing and collective action on city-level sustainable
development performance. There are also opportunities to observe more contrasting
factors in combination that causes poor city-level sustainable development as well. One
solution can be found using the fuzzy-set qualitative comparative methodology or the set-
theoretic approach (Fiss, 2011; Misangyi, Greckhamer, Furnari, Fiss, Crilly, & Aguilera,
2017; Ragin, 1987; Rihoux & Ragin, 2008). The QCA methods developed and improved
by Ragin (2014), Fiss (2011), and Misangyi et al. (2017) combine qualitative case-
oriented research to account for the complexities of city-level integrative ecosystem
management to foster sustainable innovation and development with scalable
computational methods that allow generalizing city level SDG score improvements as
well as the cities stagnation/declination linked to sustainable development performance
outcomes. The benefit of using a QCA is that I can capture large patterns occurring in my
phenomena and validate and predict outcomes with a small sample of cities to analyze.
This was my primary motivation for choosing this type of method. Another advantage of

QCA, over other methods such as generalized linear modeling (GLM), is that it is capable

of accounting for non-linear complexities (Rihoux & Ragin, 2008). The qualitative

comparative analysis (QCA) leads to holistically examine how factors combine to

produce an outcome. This study proposes the use of the fuzzy-set QCA (fsQCA) to

investigate the causes of City SDGs performance through 21 sustainable development

cases by a variety of cities and city SDG performance scores within the United States. I

aim to determine which factors, in combination, lead to high sustainable development

performance and the factors leading to low or no performance outcomes in city

ecosystems. I expect the results to provide recommendations for planning, organizing,

strategizing, and implementing sustainable development projects with high organizational

performance in cities, while also improving city-level SDG scores, which indicate a city's

progress towards flourishing cities as sustainable ecosystems (CASE). Thus, the research

question I ask is:

> *What combination of factors predicts high or low performance measures*
> *of cities working to achieve sustainable development goals?*

**Research Design**

> ***Research content.*** Twenty-one U.S. cities and sustainable development initiatives

(i.e., cases) were selected to ensure variability. Yet, I sought similarity amongst the

outcomes and factors (i.e., causal conditions). Specific case details and additional

selection method details are included in Appendix K. Major variations between cases

included implementing organizations and approaches to sustainability, municipal GDP,

regional industries, number of patents, social equity and upward mobility indicators, level

of sustainable transportation utility, and number of (R1 or R2) universities in a 50-mile

radius of the metropolitan statistical area (MSA). To control the number of causal

166

conditions analyzed, I selected each case based on significant amounts of similarities as well. Similarities included evidence of an office of sustainability or equivalent local municipal agency, presence of global industry, an urban center with adjacent suburban communities, a diversity of socio-economic and cultural demographics, established built environments, and a minimum population over 1 million within the MSA. As an additional selection process, I chose four cities from which I previously collected interviews, then selected additional cities grouped into seven segments from best to worse-performing total city index scores from the 2017-2019 UN SDG City Reports (Espey, Dahmm, & Manderino, 2018; Lynch, 2019; Prakash et al., 2017). Figure 11 displays the steps of the QCA process. Following Figure 11, I will describe my research process of data collecting, identifying causal conditions, analyzing the data in fsQCA software, and interpreting the results.

**Figure 11: Research Design, Overview, and Outcomes**

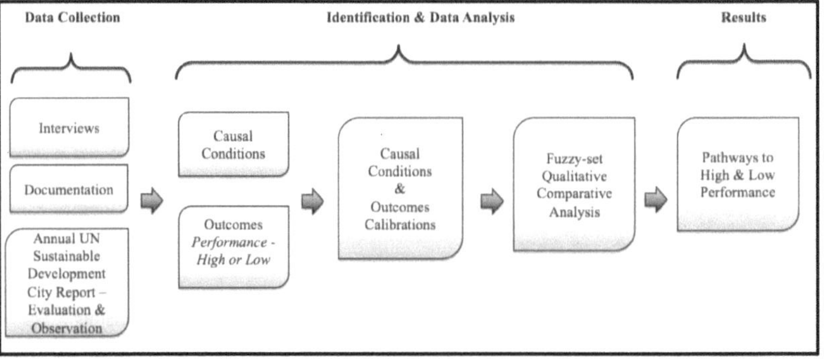


***Data collection.*** I collected a plethora of empirical evidence from previously conducted interviews, secondary data documentation from academic and in practice case analyses, and evaluation and monitoring reports for city-level sustainable development

167

performance tracking. The data were observed and read thoroughly to gain an

understanding of causal mechanisms for each city ecosystem's outcome and to generate

case knowledge that was calibrated to reveal performance outcomes. See Figure 11

above. All data were collected between July 2017 and December 2020.

*Interviews.* Semi-structured interviews were conducted with local city members

as business leaders, community leaders, local government officials, social activists, non-

profit/non-governmental organization (NGOs) leaders, participating citizens, social

organizers, and local municipality workers. Interviews explored sustainable development

implementation success and challenges from their historical experiences, current projects

in production, and future anticipated SD projects. I utilized the interview data previously

collected between July 2017 to June 2018 from interviewees that participated in city level

sustainable development initiatives within their respective cities. Participants were

selected through my professional network and consisted of sustainable development

managers, chief innovation officers in corporations, environmentalists, political officials,

university professors, municipal sustainability managers, social activists, NGO managers,

banking and investment managers, and community developers. A total of 21 interviews

were conducted to capture the lived experiences and perspectives of technical and

indigenous knowledge contributors. Participants also varied by age, demographics,

gender, and community representation. Examples of interview questions can be reviewed

in Appendix B. It was concluded that theoretic saturation was reached when no new

themes or topics emerged from subsequent interviews. All interviews were conducted in

English.

***Documentation.*** Along with the interviews that captured the "lived experiences"

of stakeholders, secondary data documentations were collected, including the UN SDG

U.S. Sustainable City Index and Rating System Reports from 2017 to 2019. This was

primarily used as a comprehensive evaluation and observation tool that highlighted the

SDGs, 57 indicators related to 15 of the 17 SD goals by a city benchmark. The SDG

Sustainable City Report is an important guide for the management of progress towards

achieving SDG goals because it highlights localized indicators for city stakeholders to

view and integrate into their planning, prioritizing, strategizing, and approaching the

challenges of transforming cities as sustainable ecosystems. One hundred five cities were

ranked by a total index score from best to worst. These total city scores were based on the

cumulative 57 indictor values of each city. Examples of key city indictors included social

dimensions of healthcare access, gender, and racial equity, environmental dimensions of

access to clean water and pollution reduction, and economic dimensions of innovation,

number of patents, municipal GDP, public infrastructure, women and minorities in

business and government, and upward mobility. Although I collected UN SDG reports—

authored by Jeffery Sachs and others in the Sustainable Development Solutions Network

(SDSN)—from 2015 to 2019, the localized indicator data underpinning each of the 17

goals were not comprehensively developed and collected on city levels until the 2017

report. Therefore, I collected the three total score data of the selected city cases between

2017 and 2019 and determined the positive, negative, or stagnate performance measures

by score changes from 2017 to 2019.

Each city's comprehensive development plans, sustainability plans, and municipal

progress reports were also collected as secondary data by web search. In addition to the comprehensive sustainability planning documents, I subscribed to various sustainability-focused programs, sustainability learning and awareness campaigns, and other community outreach initiatives and collected these observations as "memos" to stay current with any new developments uniquely relevant to each cities' sustainable development progress.

I also performed a synthesized literature search in various combinations using EBSCO. To draw upon the empirical evidence and extract related themes of sustainable development, cities, ecosystems sustainable performance, I searched using the following terms: "sustainable development," "community development," "asset-based community development," "ecosystem performance," "ecosystem development," "sustainable development goals," "sustainability," "integrative management," "macromanagement," "universities," "city development," "systems management," "political economy," "community participation," "stakeholder attitudes and politics," "innovation ecosystems," and "ecosystems management."

***Data analysis.*** Interview transcriptions, observations, memos, and documentation were uploaded into NVivo qualitative coding software. Qualitative data were coded using both deductive and inductive methods (Saldaña, 2021). For the deductive coding, I used theory to identify the important themes related to community, ecosystem, and city sustainable development of high and low-performance outcomes. For example, the themes of meta-organization for sustainability, government intervention in localized sustainable development, community participation were all identified prior to the start of coding because the literature suggested that these are performance increasing factors towards achieving sustainable development goals in city ecosystems (Berkowitz, 2018;

Dunphy, 2000; Florini & Pauli, 2018; Rousseau et al., 2019; Sachs et al., 2019; Saha, 2009; Voegtlin & Scherer, 2017).

In deductive coding, I first analyzed the data by separating themes into positive sentiments and negative sentiments in NVivo. Next, I performed open and selective coding for the emergence of new themes related to regional networks and city sustainable development performance. For example, the theme "Needs/deficiency-based community development (NBCD)" was identified during selective coding because it emerged from interviews as a community development approach that was less than successfully implemented and not yielding a positive overall community impact or performance. NBCD was also confirmed within the literature as the traditional approach to social community development where social services focus on learning and providing the needs of the community, in contrast to community development that learns and develops the community with the community (Nel, 2018). Any conflicting statements and themes between interviewees were resolved by triangulating responses with documentation, memo reviews, and other secondary data (Basurto & Speer, 2012).

*Fuzzy set QCA.* For my analysis, I used the fuzzy logic variant of QCA (Fiss, 2011; Ragin, 2014) because I found it to be a useful analytical tool for small-N samples and when the unique cases do not dichotomously fall fully "in" or fully "out" of a set. For example, local sustainable development implementation involves state government support to varying extents. Thematic sets for my causal conditions and outcomes were defined through calibration and measured by the four-value fuzzy set as membership scores (Fiss, 2011; Greckhamer, Furnari, Fiss, & Aguilera, 2018). The four-value fuzzy set ensures the set definitions provide consistent measurement calibration for meaningful cases differentiation (Basurto & Speer, 2012). I analyzed the data by constructing a "truth table" for analyzing set relationships between attributes and outcomes. This step was conducted and computed with the fsQCA software, which uses Boolean algebra to reduce the truth table to a minimum set of logical expressions that cover all logical combinations with the same outcome (high performance or low performance) solutions. The solutions outputted the integration of counterfactual conditions and measures of fit (consistency and coverage) for result evaluation. Two tables were evaluated and reported, one with configurations for city sustainable development with high performance and one with configurations for city sustainable development with low performance.

***Domain and causal conditions identification.*** Causal conditions are the hypothesized factors in sustainable development activities that influence city sustainable development performance outcomes. I identified the causal conditions from a collection of my theories, practitioner reports, and case knowledge. After identifying these factors, I initially reviewed a total of 25 causal conditions for all 21 city ecosystem cases. These were then reduced to a final set of seven causal conditions. The initial list was developed from the literature and from the UN SGG Report. One example of a selected causal condition was *leadership asset orchestration.* Both Schoemaker et al. (2018) and Heaton et al. (2019) concurred that when certain elements of an ecosystem are present yet not performing acceptably, resources must be orchestrated by a strong, powerful actor, or actors, with clout willing to take the lead; and they usually arise as a mayor or governor, a corporate executive at a local industry enterprise, or a campus leader in academia.

The initial list was then separated into two sections, one with causal conditions from the literature (Table 12) and another list containing the SDG indicators linked to social mobility, economic growth, and climate change management performance (Appendix K). Initially, I identified some of the SDG indicators as causal conditions, but after a review, I considered some to be more aligned as domain conditions. Domain conditions are those that do not vary across cases; thus, they were removed from the analysis and used as references of case characteristics (Ragin, 2014). The domain conditions included: cities with MSA populations < 1 million, multibillion-dollar GDP with GDP growth, at least one R1 or R2 academic institution in a 50-mile radius, evidence of an office of sustainability or a municipal department dedicated to sustainable development initiatives, public transit, and a mix of population demographics containing a diversity of cultures and ethnicities.

I did identify additional causal conditions from case knowledge within my interviews. Interviews uncovered *financial capacity building* and *government intervention and barriers* because some city ecosystems could not perform certain planned sustainability initiatives due to financial deficiencies and raising enough capital sufficient enough to continue and complete certain projects. Moreover, non-supportive state and national-level government agendas and initiatives not aligned with sustainability principles can leave city ecosystems in static mode of performing sustainable development initiatives. After triangulating the *government intervention barriers* causal condition, I found that Frantzeskaki (2019) concurred this to be a true barrier as it relates to city networks and urban institutional development for sustainability.

As a final step, I evaluated the causal conditions list to remove any conditions with low necessity scores (i.e., less than 0.2 is a conventional cutoff for condition inclusion (Opdyke, Javernick-Will, & Koschmann, 2018). If causal conditions were too similar, I combined them into a single condition. Lastly, if the conditions were found to be least important during the QCA minimization process, then I removed them as well.

**Table 12: Causal Conditions and Supporting Literature**

| Causal Conditions | Citation from Literature |
|---|---|
| **Comprehensive Sustainable Development Plan** | (Cooper et al., 2009; Saha, 2009; Yigitcanlar et al., 2015; Yigitcanlar & Kamruzzaman, 2015) *Case knowledge* |
| **A Meta-organizational Platform** | (Berkowitz, 2018; Berkowitz et al., 2017; Dale et al., 2010; Heaton et al., 2019; Lynch, 2019; Silvestre & Țîrcă, 2019; Westley & McGowan, 2017) *Case knowledge* |
| **Government Policy Intervention, Support, & Barrier** | (Dunphy, 2000; Florini & Pauli, 2018; Lee & Petts, 2013; Prugh et al., 2000; Rousseau et al., 2019; Sachs et al., 2019; Saha, 2009; Voegtlin & Scherer, 2017) *Case knowledge* |
| **Local University Participation in Planning, Developing, and Implementation** | (Albareda-Tiana, Vidal-Raméntol, & Fernández-Morilla, 2018; Heaton et al., 2019; Neary & Osborne, 2018; Sachs et al., 2019; Teece, 2018b) *Case knowledge* |
| **Powerful Leadership and Asset Orchestration** | (Bec et al., 2019; Heaton et al., 2019; Mousavi et al., 2018; Rant, 2020; Saha, 2009; Schoemaker et al., 2018; Teece, 2007; Wang et al., 2012) *Case knowledge* |
| **City Citizens Climate Awareness, Culture, and Attitude on Sustainability** | (Bain et al., 2019; Carley & Smith, 2013; Duxbury et al., 2012; Portney, 2005; Prugh et al., 2000; Rousseau et al., 2019; Sachs et al., 2019; Wang et al., 2012; Watene & Yap, 2015) *Case knowledge* |
| **Financial Capacity Building** | (Fisher, Geenen, Jurcevic, McClintock, & Davis, 2009; Heaton et al., 2019; McKnight, 2017; Nel, 2015; Saha, 2009; Wang et al., 2012) |

***Causal condition calibration.*** My completed calibration process for the causal conditions can be found in Appendix L. Six of the seven causal conditions were mostly founded on qualitative data, so I calibrated these conditions indirectly as set memberships utilizing the four-value fuzzy logic scoring. I set the anchor points for fully in-set membership (fuzzy set score of 1), fully out-set membership (fuzzy set score 0), and the crossover point (fuzzy set score 0.5) for each causal condition based on theory. Partial

174

membership scores for mostly but not fully in-set (fuzzy set score 0.66) and more or less

out-set (fuzzy set score 0.33) were included to complete the 4-value range of membership

points.

I then adjusted these definitions until meaningful differences between the 21 cases

were precisely reflected in their unique calibration. For example, fully in-set membership

for *Local University Participation in Planning and Developing* was when the local

university was a primary stakeholder and asset orchestrator in the planning, strategizing,

and implementation stages of sustainable development initiatives within the city

ecosystem. This goes beyond conventional university roles of just developing skills,

competencies, and training for the community. The local university provided faculty with

varying expertise to represent the university as facilitating enablers, action researcher,

and leading experiments for sustainable development initiatives in the ecosystem.

University faculty are also instrumental in offering platforms and forums within the

community to accelerate knowledge and develop dynamic capabilities to foster

sustainable innovation.

One causal condition was calibrated directly, as I defined set membership scores

by normalizing the raw quantitative data within the boundaries of my anchor points (Fiss,

2011; Misangyi et al., 2017). For example, *Local Stakeholders/Citizens awareness and*

*attitude toward climate change and sustainable development* fully in-set membership

scores were normalized from raw data and extracted from the UN SDG City Report

indicator 13.3 - *Citizen climate, culture, and awareness* scores which are collected

annually by the Yale Climate Opinion Maps, Yale Program on Climate Change

Communication[4] project (Lynch, 2019). Supplemental data and charts on the direct

calibration step can be found in Appendix L.

*Outcome classification and calibration.* Each city case's sustainable

development performance was classified as being either high performing or low

performing. Performance was indirectly calibrated by utilizing the 4-value fuzzy set

(Appendix M). High performance was defined by the overall 3-yr. change value of UN

SDG City Report total score increased or improved. Cases were classified as low

performing if their SDG total score 3-yr. change value was 1 or less. In-set membership

was calibrated indirectly as total score value improvements of 10 or greater. Out-of-set

membership was defined by total score change values of 1 and includes negative values,

which also indicates declining SD performance. Partial in-set membership was defined by

3-yr. total score increased between 6-to-9-unit points. Partial out-of-set membership was

defined by 3-yr. total score increased between 2-to-5-unit points. The outcome of

low/poor performance was analyzed using the absence of high or moderate performance

outcome scores.

*Pathway identification, truth table, and interpretation.* For all of the causal

conditions and outcomes, I assigned fuzzy-set scores for the 21 cases and summarized

results in the QCA truth table below (Table 12). I computed my data in the fsQCA add-in

package for Microsoft Excel (Ragin, 2014) to minimize the truth table utilizing the

Boolean algebra and fuzzy logic (Rihoux & Ragin, 2008) and to determine pathways. I

assessed the usefulness of pathways using the QCA metrics: consistency, coverage,

necessity, and sufficiency. Consistency measures the degree to which cases with a given

---

set of conditions exhibit the outcome. Consistency also evaluates each emerging

pathway's reliability, which represents the fraction of cases that exhibit the same pathway

and outcomes, where a consistency score of 0.7 is required to be "consistent" (Ragin,

2006). Coverage helps evaluate the generalizability of cases with the same outcome

representing the fraction explained by the same pathway (Rihoux & Ragin, 2008). Higher

coverage indicates that the pathway explains more cases. Necessity measures how

commonly a causal condition is present with an outcome. Fractions above 0.9 are

required to "necessary" (Ragin, 2014). Sufficiency measures how commonly a causal

condition results in positive outcomes; and a requirement of fractions above 0.4 indicates

a causal condition to be "sufficient" (Ragin, 2014).

**Results**

The truth table with causal conditions and outcome fuzzy scores can be viewed in

Table 13. The truth table summarizes the direct and indirect fuzzy scores for all 21 city

cases. The computation in fsQCA software revealed that there were four pathways for

high sustainable development performance (Figure 12A) where each revealed an

alternative path towards achieving the same total improvement outcomes of performance.

A total of four city cases had high performance and improved their SDG total city scores

ten points or greater. All of the four high-performing city cases shared the five causal

conditions of *Multi-level Government Intervention/Support; Financial Capacity*

*Building; High Citizen Climate, Awareness, Attitude, Culture; Leadership Asset*

*Orchestration; and structuring a Meta-Organization Platform.* Of the four pathways, one

city case, San Francisco, CA, resulted in its own unique pathway, which was

## Table 13: Truth Table and Summary of Fuzzy Score Results

| CaseID | Causal Conditions | | | | | | | Outcomes |
|---|---|---|---|---|---|---|---|---|
| | Local Adaption Plan | Meta-org Platform | Gov_Intrvtn | Leadr_Ass_Orchestrtn | Financial capacity | Univ_Part | City_Cltr_Poltic_Attd | *3 yr performance score change |
| San Francisco, CA | 1 | 0.66 | 1 | 1 | 1 | 1 | 1 | 1 |
| Washington DC | 1 | 0.66 | 1 | 1 | 1 | 1 | 1 | 1 |
| Austin, TX | 1 | 0.66 | 1 | 0.66 | 1 | 1 | 1 | 0.66 |
| Denver, CO | 1 | 0.33 | 0.33 | 0.33 | 1 | 1 | 0.66 | 0.66 |
| Salt Lake City, UT | 1 | 0.33 | 0.66 | 0.33 | 1 | 1 | 0.33 | 0.66 |
| Chicago, IL | 1 | 0.33 | 0.66 | 0.33 | 1 | 0.66 | 1 | 1 |
| Pittsburg, PA | 1 | 0.33 | 0.33 | 0.33 | 0.66 | 0.33 | 0.33 | 0.33 |
| Phildelphia, PA | 1 | 0.33 | 0.33 | 0.33 | 0.66 | 0.66 | 0.66 | 1 |
| Charlotte, NC | 1 | 0.66 | 0.33 | 0.33 | 0.33 | 0.66 | 0.33 | 0.66 |
| Kansas City, MO | 0 | 0 | 0 | 0 | 0.33 | 0.33 | 0.33 | 0.66 |
| Atlanta, GA | 0 | 0 | 0.33 | 0.33 | 0.33 | 0.33 | 0.33 | 0.66 |
| Nashville, TN | 0 | 0 | 0.33 | 0 | 0.33 | 0.33 | 0 | 0.33 |
| Detroit, MI | 1 | 0 | 0.66 | 0 | 0.33 | 0.33 | 0.33 | 1 |
| Cleveland, OH | 1 | 0.66 | 0.66 | 0.66 | 0.33 | 0.66 | 0.66 | 1 |
| Jacksonville, FL | 1 | 0 | 0 | 0 | 0.33 | 0 | 0 | 0.66 |
| St. Louis, MO | 0 | 0 | 0 | 0 | 0.33 | 0.33 | 0.33 | 0.33 |
| Houston, TX | 1 | 0.33 | 0.33 | 0 | 0.33 | 0.33 | 0.33 | 0 |
| Las Vegas, NV | 0 | 0 | 0 | 0 | 0.33 | 0.33 | 0.66 | 0 |
| Oklahoma City, OK | 0 | 0 | 0.33 | 0 | 0.33 | 0.33 | 0 | 0 |
| New Orleans, LA | 1 | 0 | 0.33 | 0 | 0.33 | 0.33 | 0.33 | 0.66 |
| Memphis, TN | 1 | 0 | 0.33 | 0 | 0.33 | 0.33 | 0.33 | 0 |

## Figure 12: Configurational Results of Sustainable Development Performance (A-high | B-low)

### A) High City Sustainable Development Performance



| Condition | Necessity |
|---|---|
| Local Adaption Plan | 0.83 |
| Financial Capacity Building | 0.73 |
| University Partartcipation | 0.73 |
| Local Citizen Climate Awareness and Attitude | 0.68 |
| Multi-level Government Intervention | 0.62 |
| Leadership Asset Orchestration | 0.46 |
| Meta-organization Platform | 0.4 |

*Solution Consistency* = 0.95
*Solution Coverage* = 0.46

* Underline indicates an explanation by >1 pathway
*Bold Italics* uniquely indicates an explanation by 1 pathway

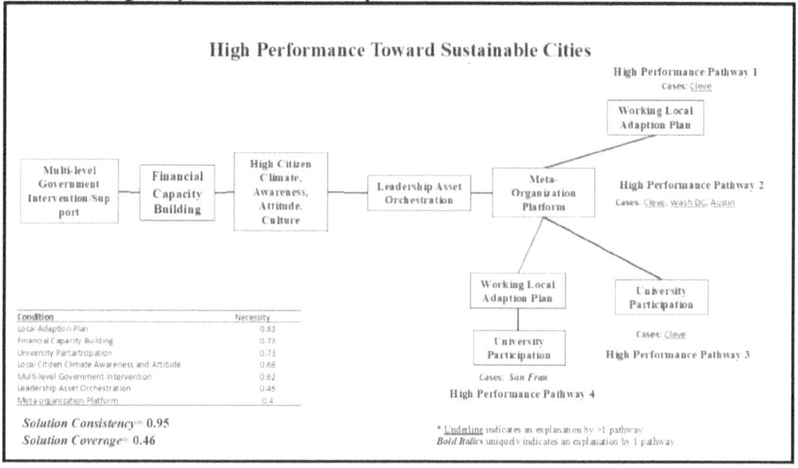

B) No to Low City Sustainable Development Performance

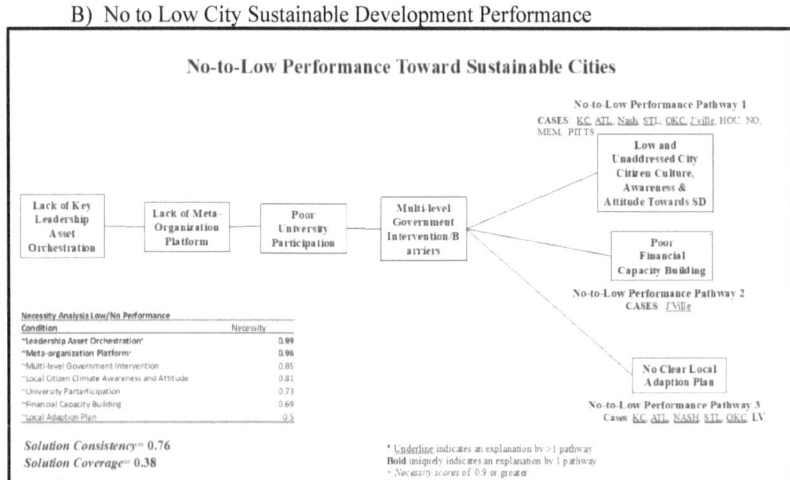



Also, the only case pathway including all seven casual conditions was San Francisco. The city of Cleveland, OH, had multiple pathways that included five to six causal conditions. The other two pathways explaining Cleveland adds a path that includes a choice of *Working Local Adaption Plan* or *University Participation* over the five causal conditions shared by the other high-performing city cases. Although none of the causal conditions resulted as a "necessary condition" (necessity scores of 0.9 or greater), all seven of the conditions resulted in satisfying the threshold to be considered a sufficient condition. This was confirmed by the necessity and sufficiency metrics analyzed by XY plotting and necessary condition analysis in the fsQCA software. The four high-performance pathways highlight the importance of a meta-organizing governance structure that is inclusive and brings all city stakeholders into the development of community and city level development and core leadership that enables and drives the

179

sharing of resources for co-designing the city ecosystem. The pathways also highlight the importance of building financial capacity from various sources, including NGOs, government agencies, philanthropists, and committed businesses. It was expected to see that having higher percentages of local citizens with climate change awareness as well as positive attitudes and cultural connections towards sustainability was a sufficient condition towards greater performance measures in the city ecosystem. Along with a comprehensive climate adaption plan to work from and/or local universities' participating in sustainable development efforts, pathways to higher sustainable development performance can be realized for the cities' shared goals of a flourishing ecosystem.

As seen in Figure 12B, there were three pathways leading to stagnation and declining sustainable development performance. The eleven poor-performing city cases shared four common causal conditions: *Lack of Powerful Core Leadership Asset Orchestration, No Meta-Organization Structure/Platform, Poor University Participation,* and *Multi-level Government Lack of Support/Intervening Government Barriers.* Low/No performance pathway 1 described ten of the eleven poor-performing city cases and included the causal condition *Low and Unaddressed City Citizen Culture, Awareness & Attitude Towards SD.* In addition to the four shared conditions, Low/No Pathway 2 included the *Poor Financial Capacity Building* causal condition that described the Jacksonville, FL, city case. Pathway 3 included the condition of No *Clear Local Adaption Plan* in addition to the four shared conditions, which described the poor performing city cases of Kansas City, Atlanta, Nashville, St. Louis, Oklahoma City, and Las Vegas. Overall, the poor-performing pathways each highlighted that sustainable development initiatives in these city cases were unable to significantly improve with low stakeholder participation and forged partnerships for sharing resources, building capital, and developing new capabilities. In addition, a lack of a whole systems governance structure and key leadership resource orchestration and financial capacity building with no comprehensive adaption plan is difficult to overcome; especially when there is a low consensus within the city population that lacks climate change awareness and are inactive and uninvolved with the local sustainability movement.

Solution consistencies were moderately high for both high and low/no performance pathways; however, coverage measurements for them both were moderate at best; which indicted relatively strong results. Additionally, my pathways represented the combinations of causal conditions that together were sufficient to produce outcomes of poor and high-performance measures. In Figures 12A and 12B, I displayed my causal conditions in the order of decreasing necessity scores. Although I had no necessary conditions in my high-performing pathways, I did have two causal conditions, *Lack of Powerful Core Leadership Asset Orchestration* (0.99) and *No Meta-Organization Structure/Platform* (0.96), in my low to no performance pathways that scored as a necessary condition. My results exhibit alternative combinations of conditions that lead to either high or low sustainable development performance in city ecosystems. Regardless of the outcome, each pathway was sufficient to produce that particular performance outcome. It is important to note that I am not advocating that any one pathway is better or worse than another, but I do express that there are multiple pathways city ecosystem's implementing organizations can take to achieve a high-performing sustainable development program that leads to flourishing cities as sustainable ecosystems.

**Discussion**

*Pathways to high performance.* I identified five factors shared by city ecosystems that led to pathways of high sustainable development performance: financial capacity building, powerful leadership, asset orchestration, meta-organization formation and platforms, and multi-level government support. Although none of my "high performing" factors resulted as "necessary," necessity scores revealed that, at minimum, all factors were sufficient in the final configural model. The final configural model for high performance was relatively strong based on the results of the evaluation metrics (consistency, coverage, necessity, sufficiency). My configural model was not developed sequentially or in chronological order but was developed using a necessary and sufficiency order based on the fsQCA's Boolean algebraic results. Since all of my factors, at best, are "good to have" for my configural model, I will next observe each factor and its influence in this configuration on my high-performing city cases from the study.

Adequate policy and financial support from national and state agencies are important to influencing city and municipal sustainable development efforts and addressing climate change in our cities' infrastructure. However, due to constraints in project budgets, diminishing tax revenues, and political gridlock, contemporary community development projects and infrastructure upgrades intensify more challenges to most city-level sustainable development progress as they search for more resources and sustainability-focused policies from higher government agencies to support their efforts.

A city ecosystem has a self-organizing platform to operationalize the shared vision and goals of flourishing. It can start with moderate partnering and formation of organizations and institutions, but the full spectrum of organizations, stakeholders, industry sectors, and community representation to experience the highest performance anticipated. The ecosystem organization functions by heterarchy making decisions on consensus, including heterogeneous actors that facilitates knowledge transfers, dialogue, and collective learning. Multi-stakeholders within the implementing organizations produce reporting mechanisms to enhance social responsibility and sustainable development, focusing on transparency and accountability of all stakeholders within the ecosystem. Innovation is consistently developed for social benefits across the city ecosystem, and information is produced to develop more capabilities and social innovation. This is probably why government support and intervention resulted as "not sufficient, but good to have" because all of my city cases experienced deficiencies in higher-level government support as well as their own budget constraints to implement sustainable infrastructure projects. Yet, the cities of San Francisco, CA; Cleveland, OH; Washington, DC; and Austin, TX were able to holistically improve each of their own SDG total scores, despite varying government support on the state level and minimal support on the federal level. For example, the state of California has been a vanguard in sustainable development initiatives, policymaking, and moderately supporting its cities sustainable development for almost half a century, while southern U.S. states, including Texas, have been slow to seriously adapt and significantly support sustainable development and addressing climate change up until the last decades or so. These four cities' high-performing quality over the three-year period, which also reveals the progression towards achieving sustainable city ecosystems. In the case of Cleveland, I found an interesting example of higher-level government intervention. Extracted from my interviews and secondary data, I found that the State of Ohio had its own energy-focused sustainability plan that did not align well with the Cleveland Northeast Ohio's plan. By 2050, the City of Cleveland and Cuyahoga County region's "Sustainable Cleveland - Green City on a Blue Lake" plan has a goal of attaining 100% clean energy, which would include an integrated energy system consisting of a wind turbine farm on Lake Erie, solar energy program for residents and businesses, natural gas, and a novel energy storage system that provides clean energy for their citizens and industries. However, the state's energy plan had a heavy emphasis on coal and oil infrastructure in place for the rest of the state, probably due to substantial influences of those energy industries still existing in the southeast region of the state. Since the Cleveland ecosystem is moving forward with its plan, this demonstrates an example of a city ecosystem taking charge of their communities and progressing towards a sustainable city without strong support from their state.

Financial capacity building has always been the cornerstone of capitalistic societies, and its relativity to sustainable development is critical for high performance. However, in sustainable development projects, conventional methods of financial capacity building are being supplemented or replaced through innovative financial means. For example, communities are now pooling resources available within their own philanthropic endeavors, businesses organizations, and public-private partnerships together, sharing risks and benefits. This has also been a response to deficit-strapped governments when proactive communities and city implementing organizations are committed to their sustainable development performance but do not have the full

financial capacity to operate. Building capital is an important factor towards high

sustainable development performance, but it is not a requirement for success.

Powerful leadership asset orchestration is considered a key dynamic capability for

ecosystem management, developing sustainable innovations, and sustainable

development (Heaton et al., 2019; Mousavi et al., 2018). Powerful leadership asset

orchestration is the catalyst in the initial stages of forming the implementing meta-

organization, promoting the shared vision for the city, enabling more partnerships across

sectors, coordinating resources, and extending decision-making and co-designing power

amongst participating stakeholders in the city ecosystem (Linde et al., 2021; Mousavi et

al., 2018). I believe that power leadership asset orchestration to be a required factor;

however, since it was evident in all high-performing cities, I consider it an essential

factor for predicting high sustainable development performance. At the highest level of

performance, powerful leadership asset orchestration is defined as the full formation of

ecosystem leadership where powerful actors initiate a shared city goal of flourishing,

sense and seize opportunities to strengthen city ecosystem's resilience, expand the

leadership base by bringing in other city stakeholders, share the managing and decision-

making across sectors and social systems, while also organizing organizations,

institutions, and citizens; integrates and coordinates stakeholders roles and promotes

innovation towards better resource utilization for the city and its stakeholders for the

reconfiguration toward flourishing and sustainable cities. Examples of this factor were

found in some of my cases and were evident in the San Francisco/Bay Area with the

leadership taken charge from UC Berkeley's former chancellor Nicholas Dirks innovation ecosystem structuring, Stanford University and CEOs of Silicon Valley, and most of the city and suburban municipal governments partnering together (Heaton et al., 2019). SkyDeck, a popular start-up accelerator, developed as a collaboration between UC Berkeley's engineering and business school with the local communities to anchor a pipeline for local commercial innovation development was a more specific program that gained my attention because they worked heavily in sync with political leaders, city government officials, and venture capitalists in the region (Oh, Phillips, Park, & Lee, 2016). I also found this high level of collaboration and orchestration between the city of Austin, local industry leaders, and the University of Texas-Austin. The City of Cleveland, led by mayor Frank Jackson, orchestrated a program called Sustainable Cleveland where local community, business, and thought leaders formed a partnership to approach sustainable development goals in Cleveland. The Cleveland example was unique because the Sustainable Cleveland program was facilitated by Dr. David Cooperrider and Case Western Reserve University, which also utilized Appreciative Inquiry in the design process and co-creation capability for the city-wide sustainable development program.

The Cleveland and San Francisco cases resulted in at least one unique pathway. All of the factors were evident in San Francisco, which is probably why they are the top sustainable development performing city ecosystem and has consistently been ranked the #1 ranked sustainable city of the U.S. Cleveland was the other case that had a unique pathway, which included the five shared factors, and accordingly, the factor university participation.

The degree of university participation in local or regional sustainable planning and implementation and quality of engagement of the university with the local community was an important factor for high sustainable development performance. At the highest level of participation, universities that play primary stakeholder and asset orchestrating roles in sustainable development planning and co-designing and go beyond conventional roles of just developing skills, competencies, and training for people as a primary facilitator of the community development process. I found better than average university participation examples, as I mentioned in the Cleveland, San Francisco, and Austin cases; however, only the San Francisco and Cleveland ecosystems resulted with this factor in their pathway.

From my analyses, it was interesting reviewing annual comprehensive city and sustainability adaption plans. From the 21 cases, all city ecosystems had a "comprehensive" development plan; however, there were varying levels of sustainability principles and development projects planned or in the implementation phase included in the document. There were also certain cases where comprehensive city plans were separate documents from sustainability and climate adaption plans. I chose to include the raw data from the Adaptation Clearinghouse, Georgetown Climate Center research project, which only reported in binary values of having a plan or not. This may have affected my configural model results, and I caution that a more aligned and varying calibration of set membership scores may be warranted for future research.

Local citizen's climate, culture, and awareness reflect the estimated percentage of adults who think global warming is happening, and human intervention is needed to address it in their respective city ecosystem. It was a factor that had a significant influence on high sustainable development performance in all of my high-performing city cases. This factor was directly calibrated for my study, but it was interesting to see that

187

the raw scores were heavily varied, and they did not correlate well with the raw total city scores associated with that city. For example, as with most southern U.S. cities, New Orleans has consistently been one of the lowest rank U.S. cities as it relates to total city scores and achieving SDG goals, according to the UN SDG city report. Yet, the citizen climate awareness, attitude, and culture indicator score for New Orleans's 2019 report indicated that 42%, or close to half of the population in the area, considers climate change as a problem and needs to be addressed. In comparison, neighboring cities ecosystems, such as Jackson, MS, and Baton Rouge, LA, which are the lowest-ranked cities, have citizen climate, culture, and awareness scores of 15% and 18%, respectively. Top ranking San Francisco/Bay Area capped the citizen climate awareness, attitude, and culture indicator score at 100%, but conflicting indications revealed top-ranking Salt Lake City scored only 39%. This exemplified the non-correlation between total city scores and citizen climate awareness, attitude indicators to achieving sustainable development, but it also indicates the relative importance of the localize social ethos and its relative factor importance in high-performing pathways to catalyze action and high-performance in city ecosystems towards their total sustainability goals. These findings and data discussed can be reviewed in the 2019 US Cities Sustainable Development Report SDG index table in Appendix K.

*Pathways to poor performance.* There were 11 of the 21 city cases that revealed the three pathways to poor performance: Kansas City, Atlanta, Nashville, St Louis, Oklahoma City, Jacksonville, Houston, New Orleans, Memphis, Pittsburg, and Las Vegas. I was surprised that this performance percentage was aligned closely with the three years of UN SDG city reporting, which has consistently stated that slightly over half of the 105 U.S. cities tracked in the report are stagnating or declining in their sustainable development performance and total goal attainment. Despite my study extracted about a fifth of the cities as samples and I combined organizational strategic mechanisms with selected SDG indicator raw data to find pathways to high and low performance, my results are closely similar to the 50+ SDG indictors, total SDG city score results, and other findings from the UN SDG comprehensive report.

I used the "negate" or absence of a factor in a pathway to analyze and compute low to no sustainable development performance. As a result, the lack of powerful key leadership asset orchestration and lack of meta-organization emerged as necessary factors that lead to poor performing pathways. The lack of these two factors highlights the importance and need for leaders and implementing organizations to consider heterarchical organizing and governance structuring within the community to reverse the poor performance effect, share resources, and extend decision-making opportunities to other stakeholders within the development process. Moreover, the impact of a powerful leader (or group of leaders) with a high level of clout builds trust and empowerment to action within the ecosystem to influence change and promote the innovative process.

Lacking considerable participation from local universities and poor state/national government support and policymaking are also significate impeding factors leading to poor city ecosystem performance. I found several examples of these factors in my interviews and secondary data. And although they were highly sufficient (necessity scores 0.85 and 0.79 respectively) and trending towards necessary factors, it is important for stakeholder leadership to forge relationships with local universities and encourage faculty to take a greater role in the city's sustainable development progress. Additionally,

189

it may be beneficial for local leaders and government officials to anticipate limited
support from states and consider an approach to their sustainable development progress
focusing internally on resources and support within the community.

Including the four shared "deficiency" factors of meta-organizing, university
participation, government intervention, and leader asset orchestration, low and
unaddressed climate change awareness and citizen's attitudes towards sustainability
completes the first pathway to poor performance. I found evidence of this pathway in 10
of my poor-performing cities. To address low citizen climate awareness and reverse
sustainable development performance, it may be helpful to have consistent outreach and
educational programs and ongoing campaigns incorporated into the local office of
sustainability. An example of this was found in Cleveland, where the Sustainable
Cleveland Initiative has spawned social media campaigns, held webinars and forums, has
had an annual Appreciative Inquiry Summit for 20+ years, and these strategies have been
effective in reversing the pathways positively towards high-performing sustainable
development in their city ecosystem.

The case in Jacksonville revealed another pathway to poor performance. In
addition to the four shared factors, Jacksonville had a fifth factor in the second pathway
to poor performance that was poor financial capacity building. The lack of a clear climate
adaption plan completed the third pathway to poor performance and was evident in
Kansas City, Atlanta, Nashville, St Louis, Las Vegas, and Oklahoma City; and it was a
unique pathway that explains the case in Las Vegas. The lack of comprehensive
sustainability or climate adaption plan may be an indicator of the leadership's true
commitment to achieving sustainability or may be an indicator of technical and

managerial capacity building in the ecosystem or lack of community stakeholder interest. However, having a climate adaption plan does not guarantee high city performance. For example, Memphis, Houston, and New Orleans have a working climate plan, yet these city ecosystems have resulted in pathways to low to no sustainable development performance.

Overall, if city leadership and community stakeholders are serious about reversing the effects of climate change, and desire to promote flourishing communities within their ecosystems, it is important that they first understand that conventional methods community development and deficiency focused approaches to sustainability lead to poor performance as it relates to city level sustainable development achievement. Novel approaches such as enabling informal and formal partnerships through meta-organizing, deep collaboration, designing consistent sustainability outreach campaigns, with good asset orchestration by a community leader or leaders create different pathways to high sustainable development performance.

**Conclusion**

There were four pathways that led to high city ecosystem sustainable development performance and three pathways that led to stagnation and declining sustainable development performance. The top-performing city, San Francisco, found that *Multi-level Government* Intervention/*Support; Financial Capacity Building; High Citizen Climate, Awareness, Attitude, Culture; Leadership Asset Orchestration; and structuring a Meta-Organization Platform, Working Local Adaption Plan* and *University Participation* were all sufficient in combination to organize the community and all relative stakeholders, forge partnerships across sectors, and have adequate key leadership orchestration in planning, strategizing, and implementing sustainable development goals. The other high-performing cities found it sufficient to have just *Multi-level Government Intervention/Support; Financial Capacity Building; High Citizen Climate, Awareness, Attitude, Culture; Leadership Asset Orchestration; and structuring a Meta-Organization Platform* for their high-performing measurement, while the Cleveland ecosystem additional included either *Working Local Adaption Plan* or *University* Participation for their improving performance. Overall, the pathways the high performance demonstrate the importance of being inclusive of all stakeholders in all communities of the city; utilize a meta-organizing governance device for asset orchestration, financial capacity building, and sharing resources to forge new sustainable innovation and capabilities across the ecosystem; and a clear adaption plan to continuously increase awareness and participation in the community for ecosystem change. Each of the four pathways can lead to high-performing sustainable development transition; therefore, multi-stakeholder implementation should focus on aligning and sharing available resources, knowledge, capabilities, assets internal and external to the core city ecosystem to develop sustainable value propositions to cultivate a thriving and healthy city environment.

Poor-performing city ecosystems lacked many of the important conditions for high performance, and the results demonstrate many complex causes of stagnation and declining performance measures. All poor-performing cities had a *Lack of Powerful Core Leadership Asset Orchestration, No Meta-Organization Structure/Platform, Poor University Participation,* and *Multi-level Government Lack of Support/Intervening Government Barriers;* and because a *Lack of Powerful Core Leadership Asset Orchestration* and *No Meta-Organization Structure/Platform* emerged as "necessary" conditions toward low sustainable development performance, I recommend that implementing organizations and institutions be cognizant of internal strengths, assets, and stakeholders within the community that sometimes get overlooked and underappreciated; which aligns with the nature of AI and SOAR. The pathways to low performance emphasize the importance of not only engaging multi-stakeholders, but also enable untapped resources that may be hidden or unconsidered in the design process of community development and particularly, sustainable development towards flourishing ecosystems. This way, poor-performing city ecosystems can establish clear mechanisms to deploy sustainability principles into the city ecosystem and reverse the trajectory of sustainable development performance.

## CHAPTER 8: INTEGRATED CONCLUSION

I asked the primary question of how integrative system approaches are instrumental in the development of flourishing cities as sustainable ecosystems. The combination of the three studies supports an explanation that solves the primary inquiry of my book and advances knowledge for the meta-theory of ecosystem sustainable development and how to achieve the developmental identity of sustainable and flourishing. I asked three subsequential research questions that assisted the building of my argument and provided solutions to each study and primary inquiry by a mixed-methods methodology. The research question was addressed through an open mixed methods approach (Tashakkori & Creswell, 2007; Teddlie & Tashakkori, 2006) and composed complementary studies that fit to create the multi-faceted understanding of sustainable development performance and achieving sustainability goals. I found the mixed methods approach to be appropriate for the "analysis of the different facets of city ecosystem sustainable development that yielded an enriched, elaborated understanding of this phenomenon" (Quinney et al., 2016). I provided a unique understanding of ecosystem sustainable development by positioning the three research questions and empirical designs as *complementaries*, where each of the studies focuses on a different aspect of city level sustainability and flourishing, and by *expansion* where each study extends the scope of inquiry more broadly and holistically instead of more reductive and narrow (Greene et al., 1989).

My mixed methods approach combined qualitative and quantitative studies, with the nature of the research questions and data requirements guiding the sequencing, relative emphasis, and points of integration and connection (Johnson & Onwuegbuzie,

2004). Moreover, the nested design of my complementary research questions and study helped to acknowledge the *nested* as well as the whole of ecosystem phenomena to capture and describe specific embedded phenomena, whole systems strategic approach, that underpins the city ecosystem identity and its internal dynamics that impact sustainable development performance (Teddlie & Tashakkori, 2006). Studying the meta-theoretical framing of ecosystem sustainable development integrated with the nested meta-organization, macromanagement, and dynamic capabilities frameworks was an approach for understanding emerging concerns and problem domains by building, applying, and studying the effects of integrated systems management processes for sustainable development in city ecosystem. The meta-theory of sustainable development framework, as I will further discuss, is a valid and complementary paradigm for exploring the internal-external phenomena of ecosystem management and dynamic changes they yield. In addition, the dynamic capabilities of meta-organization and macromanagement foster the transformation of CASE and the social consciousness towards the holistic organization and functioning of flourishing ecosystems.

The goal of integrating the three studies was to explain how an integrative ecosystems management approach, which incorporates meta-organizing macromanagement and the development of dynamic capabilities, was an effective combinational approach to addressing city-level sustainable development and achieving cities as sustainable ecosystems. I offered the meta-theoretical lens of integrative systems management for city-level sustainable development as a descriptive that has the potential to resolve the holism-reductivism paradox in the study of sustainable development and societal goals of flourishing. At the core of the paradox is the systematic issue of the

195

presence of the whole and its parts and the underpinning paradox of continuous growth

and exploitation while also maintaining the ecosystem's capacity within the natural

ecological boundaries of production, consumption, and waste outputs. The later paradox

refers to the collective interior functioning that influences the whole ecosystem and gives

the ecosystem its form and identity. The integrative ecosystem development lens

acknowledges the perspective of viewing the interior-exterior identities of a city as an

ecosystem. Cities, the exterior, can be viewed as having properties and characteristics of

ecological ecosystems that transition through phases of birth, growth, maintenance, and

release (deterioration) and back to rebirth. Interior to the ecosystem is the nested

interconnected panarchic systems functioning, which is in the form of collective human

activity and transactions occurring within the ecosystem. In this book, the focal activities

were the collective actions and operations of public and private organizations,

institutions, governments, and other stakeholders organizing and developing new value

propositions, innovation, and capabilities toward the achievement of sustainable and

flourishing developmental identities. I argue that the interior functions of the city

ecosystem, human and social activities, and the exterior dynamics that shape the

ecosystem's ecological and social environments are inseparable. And what this means is

any mediating or intervening managerial approach should consider the holistic view of

the sustainability challenge, as well as adopt holistic strategies that are inclusive of all

ecosystem elements and stakeholders.

All stakeholders in city ecosystems have some effect on the identity of a city by

how stakeholders collectively adapt to the ecosystem and external environments.

However, I argue that employing integrative ecosystem management is a better approach to sustainable development progress that fuses diverse actors' and organizations' efforts together for the functions of exploitation (cultivation), conservation, reorientation (coordination and integration), and reorganization (reconfiguration) in the ecosystem. This is the temporal flow of events that influence innovation and resilience of the city ecosystem by stakeholders and implementing organizations in sustainable development. Stakeholders in integrative ecosystem management also influence the transitioning of ecosystem phases, as, through time, city ecosystems are susceptible to processes of deterioration, low growth performance, and loss, as well as processes of reconfiguring available assets and resources to develop the city into a renewal phase as a sustainable and flourishing ecosystem. Conventionally, most cities in the U.S. particularly, have approached sustainable development with separated, reductive approaches, where top-down technical expertise drives the planning and execution of sustainable development goal attainment and implements change management strategies into the community. As Edwards (2009) suggested, this is equated to "doing the same thing over and over" and "focusing only on economic growth performance" by translational growth and development that is limited to achieving at best a city ecosystem that is committed to sustainable development but has yet to materialize an effort towards achieving the developmental stage and identity of sustaining. Moreover, I posit that change management on the scale of city ecosystems requires intervening factors of *plurality* to achieve an identity of a sustainable city with sustainability principles embedded within the social, political, cultural, economic, and built environments; and *integration* to transition towards a flourishing city as a sustainable ecosystem, which is a higher form of developmental identity that goes beyond traditional sustainability principles by broadening their goals towards no to very little harm to social and ecological environments, and encourages mass social prosperity and well-being. For ecosystems and stakeholders within them to achieve the sustainable or flourishing levels of sustainability involves transformational growth, which emerges from the collective consciousness of stakeholders within the ecosystem that catalyzes the radical shift needed to cross the ontological gap away from conventional separateness thinking and towards more interconnectedness thinking related to growth and sustaining the ecosystem. From the three empirical works in this book, I constructed a meta-theoretical model of integrative systems management for sustainable development built as the combination of macromanagement, meta-organization, and developing dynamic capabilities. I suggest that each of these components are the building blocks for an integrative ecosystem management approach towards achieving sustainable development goals in pursuit of a city ecosystem's identity of sustainability and on towards flourishing. In the next sections, I will discuss how each of these constructs is a unique and integral part of the integrative ecosystem management process based on the empirical evidence from my studies. I also will deliberate on what I found as the combinational pattern of these elements as it relates to high-performing metrics towards achieving flourishing cities as sustainable cities.

**The Importance of Macromanagement**

Macromanagement strategies aligns vertical and horizontal the ecosystem's human resources, or stakeholders, for positive future-focused collective action towards sustainability (Cooperrrider, 2012; Fredrickson & Losada, 2005; McFarland, 1977; Seligman et al., 2013). It neither a top-down, nor bottom-up approach; but the fusion of both approaches which encourages the inclusion of formal and informal actors in the strategic planning and implementation of sustainable development performance. I discovered evidence of macromanagement in my interviews with sustainability managers and stakeholders in Study 1. As one manager expressed:

> When we work towards [sustainable] development, we have to make sure we're innovating and generating societal benefits with an "and" …… This benefits this group, and benefits that group, and it benefits the environment, and…. so on.  (Respondent - CHIACA)

This statement elucidates generativity and positivity, which are conceptual elements associated with Appreciative Inquiry as a macromanagement platform. From the interviews in the Chicago ecosystem case, I found that they did not formally utilize Appreciative Inquiry as a macromanagement platform. However, these underpinning concepts associated with macromanagement were apparent, which I deduced was a common theme in the planning conversations for Chicago's shared vision of sustainability achievement. Study 2 highlighted my concept of macromanagement, where I created a conceptual model for understanding and measuring the effects of Appreciative Inquiry platforms with the SOAR framework (AI-SOAR) for advancing innovative capacities and building resiliency within the city ecosystem. In Study 1, I primarily conducted interviews with leaders, managers, and technical experts, which can be considered more as "top-down" perspectives towards sustainable development and achieving sustainable cities. In Study 2, I chose to broaden my sampling to capture a more diverse set of stakeholder perspectives in the process of city sustainable development achievement. With over 340 respondents from those in leadership roles to concerned citizen, I captured what I considered an appropriate mix of top-down and bottom-up representation in city ecosystems. As a result, Study 2 validated and confirmed through SEM that AI-SOAR has a positive effect on building ecosystem innovation and resiliency. By confirming AI-SOAR's positive effects on innovation and resiliency in Study 2, contributions emerged from this empirical work: (1) By quantitative analysis, I confirmed decades of empirical evidence on the effectiveness of AI-SOAR as an organizational system's change management device, and (2) By designing a structural model that is measurable to explain how AI-SOAR and this scale can be used as an instrument for ecosystem stakeholders to evaluate their city's progression in influencing and maintaining a healthy ecosystem vitality. As I introduced in the literature review, the ecosystem vitality is the basic functional properties and dynamics between innovation and resiliency of an ecosystem (Holling, 2001), and the organizational system of AI with the process of the SOAR framework as a construct was validated as an effective macromanagement approach that has the potential of advancing sustainable innovation and development which can affect the ecosystem vitality in a positive trajectory. Furthermore, I suggest that it was the underpinning concepts of macromanagement with the SOAR strategic process for planning, strategizing, and implementing that stimulated the innovation and resiliency processes. For example, in my survey, I asked stakeholders in cities and communities how they perceived their participation and others in sustainable development progress of their respective cities. There were unique survey items that focused on the underpinning concepts of positivity and generativity but also underpinning concepts of idea generation by considering the whole systems benefits, all stakeholder needs, and sharing information across systems in the ecosystem. As a result, my multi-stakeholder sample perceived that these are critical elements that influence innovation but also builds the resiliency to maintain the evolutionary finesse and vitality of the city ecosystem; thus confirming the importance of a whole systems macromanagement approach embedded with positive organizational principles and focusing the ecosystem co-design process towards the future sustainable city, instead of a design process that develops to remedy the past or what's been done before. Although I did not directly observe macromanagement in Study 3, I do acknowledge that Appreciative Inquiry platforms have some commonalities with certain attributes of meta-organization, which was a configural factor in pathways towards high sustainable development performance in Study 3's results. Additionally, AI has certain embedded characteristics of developing dynamic capabilities, such as the presence of powerful actors that catalyze and facilitate the initial stages of designing the AI platform, expands partnership formation across sectors and systems within the ecosystem, and that follows a developmental design process that draws from strengths and focuses exploiting current the internal assets and capabilities to create new ones. I will discuss these aspects further in the following sections. Overall, by connecting top-down with bottom-up approaches to promote the achievement of sustainable cities, I posit that this can lead to better sustainable development outcomes and performance. In addition, stakeholders confirmed that AI-SOAR, a social construction instrument and example of macromanagement, is an effective tool for structuring positive and generative ideas, value propositions, and innovation for social benefit, as well as a device for co-designing with prospection and stakeholders' focusing on future aspirations on the identity of their cities.

**The Importance of Meta-Organization Formation and Developing Dynamic Capabilities**

In this book, the concept of meta-organization was defined as a governance device that

structures organizing organizations and actors into a platform for knowledge pooling, decision making, measuring, evaluating, and holding collective members

accountable for creating sustainable ecosystem innovation (Berkowitz, 2018; Linde et al.,

2021). In my first study, meta-organization emerged as a factor sustainability managers

and leaders stated was an important factor that led to successful sustainable development

implementation. Although there was a consensus reached by the technical experts in

Study 1, I found varying types of meta-organization formation and coordination, as well

as varying applications of the other attributes of meta-organizing. The collective

interviews in the Charlotte case of Study 1 explained a notable example of meta-

organization. Envision Charlotte[5] was the city's platform that was created as a public-

private plus collaboration for the shared goal of making Charlotte a "Smart and

Sustainable City." One of their sustainability projects included the 10-year plan to reduce

green gasses and energy reduction in the City Center. Envision Charlotte and the

Charlotte Sustainability Office enabled other stakeholders in the city to partner and be a

part of the change process in the city. For example, the energy design challenge was

extended to include private organizations and nonprofit organizations in the community,

the energy powerhouse Duke Energy, City Center workers in private and public

buildings, and the University of North Carolina-Charlotte's Physics and Engineering

Departments. The decision making, measuring, evaluating, and designing were extended

to the internal stakeholders in formal recurring meetings, classes, and forums, and it was

clear that informal participation was also evident as people who worked in the City

Center had opportunities to provide feedback on how to save energy and reduce

contributions to greenhouse gasses.

---

Another case where I determined the presence of meta-organization was in the Cleveland case. Although there have been many cities around the globe that have used Appreciative Inquiry as a platform for community development, Cleveland was the only case in my three studies where there was the formal utility of the organizational system with the strategic process of the SOAR framework. As I found from Studies 1 and 3, the Cleveland case was on a pathway towards high sustainable development performance, and the successful implementation of sustainable development projects were associated with the Sustainable Cleveland Initiative, which has elements of macromanagement, meta-organization, and developing dynamic capabilities within this AI-SOAR grounded platform. As dynamic capabilities determined what those involved in ecosystem management were able to do and how effectively they could create change (Teece, 2018b), it was the initial stages of leadership asset orchestration that catalyzed and enabled the expansion of Sustainable Cleveland. I have recognized AI-SOAR as a macromanagement approach, but AI-SOAR as a platform is structurally designed as both a multistakeholder process and a meta-organizing device for developing dynamic capabilities. In Cleveland, this started with the formation of the core leadership in the city, the mayor as the champion, the office of sustainability as the logistical enabler, and Case Western Reserve University as the facilitator. Once this core leadership was established, the shared goal of the city was established, and initial community buy-in was extended to public, private, nonprofit organizations and foundations. Funding for the Sustainable Cleveland Initiative was secured by pooling financial resources from various entities in and outside the city, but by also fundraising and crowdsourcing methods by active citizens as well. Once more, key stakeholders in the city partnered to participate in the city-wide event; the AI steering committee planned the annual AI summit, Sustainable Cleveland, and over the last 20 years, Cleveland has transitioned from one of the worse performing and unsustainable cities, according to UN SDG 17 City Report rankings, to a high performing ecosystem on a pathway towards establishing Cleveland as a sustainable ecosystem. I deduced this from Cleveland's pathway to high performance in Study 3, and induced from Study 1 interviews as the sustainability managers described the heterarchical arrangement of the AI platform, the coordination and integration of organizations, institutions, and stakeholders, the pooling of resources and capabilities, the leadership asset orchestration in the initial stages, the expansion of decision-making unto all stakeholders, and the accountability that is vetted and aspired upon participating citizens to implement change and create new value propositions for the city.

Combining the macromanagement, meta-organization, and dynamic capabilities functional elements as an integrative ecosystem management approach creates this "continuous morphing" and reconfiguration; and illuminates the paradoxes between innovation and resilience, growth and maintenance, which advances the evolutionary fitness of city ecosystem (Holling, 2001; Rindova & Kotha, 2001; Teece, 2018b). The integration of macromanagement, meta-organization, and dynamic capabilities are the internal components that complement each other and influence developmental identities of sustainable or flourishing in individual human systems as well as the developmental identity of the external whole of the ecosystem (Edwards, 2005, 2009; Sachs et al., 2019). Thus, an integrative systems-based approach towards developing CASE cultivates ecosystem resources, for example, human, political, built, social, natural, financial, and cultural, so that the ecosystem transforms into higher levels of sustainable identity. After

204

analyzing the data collected in the three studies, I confirmed my assumption that most cities approach sustainable development by conventional reductive means. However, there are growing numbers of cities that are experimenting with whole systems and integrative ecosystems management to guide them towards sustainable and flourishing developmental identities.

To explain the benefits and effectiveness of the meta-organizing, macromanagement, and dynamic capabilities combination of integrative ecosystem management, I revealed how some of the cities in my studies showed hyper-translational growth of reductivists towards approaches reaches at best a committed to sustainable development identity, while other cities experimenting with integrative ecosystem management approaches reveal transformational city growth that crosses the ontological gap that transforms the city ecosystem holistically towards sustainable and flourishing developmental identities. As a reminder, there are currently no U.S. cities on a trajectory of achieving 2030 sustainable development goals, and the best performing city is the #1 ranked San Francisco/Bay Area. But as I reviewed the UN SDG global rankings, the U.S. is ranked #31[st,] and San Francisco is ranked #71st out of all developed and developing global sustainable cities. Figure 13 maps how cities temporally migrate between different stages and identities of development by reductive management and integrative ecosystem management approaches.

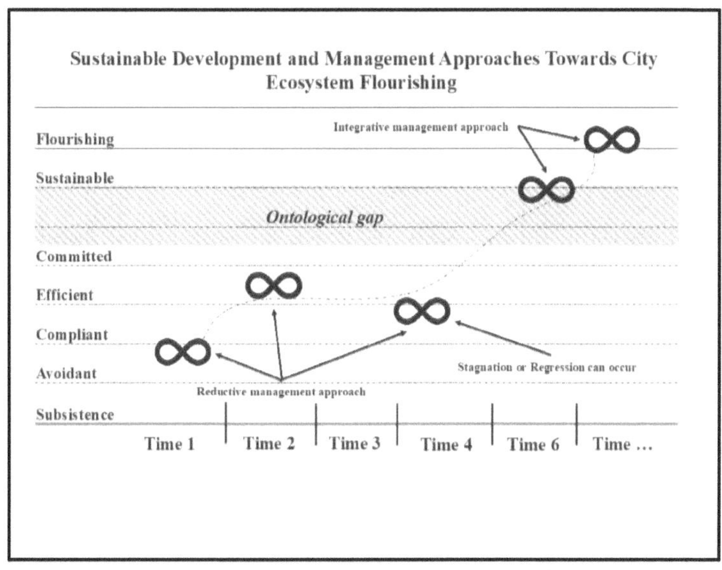


I reintroduce Figure 13 to discuss how it relates and is explained by my
complementary three studies. From Studies 1 and 3, I extracted evidence from the
Cleveland and San Francisco cases that suggest forming meta-organizational formation
and building capacities can develop dynamic capabilities during the initial formative and
planning and stages of sustainable development initiatives, such as the dynamic
capabilities developed by powerful leadership orchestration, to the recurring
implementation stages that include multi-stakeholders in the designing process, such as
coordination, seizing opportunities, and integrating systems vertically and horizontally
for holistic ecosystem participation and shared benefits. This, in turn, catalyzes the whole
design process, where the ecosystem's dynamic capabilities create new value

206

propositions to promote sustainable innovation and sustainable development within the city ecosystem. Although there is still work to do in these respective cities, city ecosystems like Cleveland, San Francisco, Washington, DC, and Austin are on pathways of crossing the ontological gap towards sustainable and flourishing identities. This was evident from my interviews in Study 1 and the pathways to high-performance configurations in Study 3. Study 2 focused on the macromanagement approach by Appreciative Inquiry platforms, and in combination with meta-organization and developing dynamic capabilities, completes the essential components of an effective integrative ecosystem management approach to sustainable and flourishing city ecosystems. The results of Studies 2 and 3 indicate that my three factors, in combination, also associated with transformative ecosystem growth, develop the capabilities for the city ecosystem to migrate from conventional stages of efficiency and committed developmental identities towards sustainable and flourishing identities (see Figure 13). However, Study 3 highlights contrasting effects that revealed pathways to poor sustainable development performance and the combination of lacking factors that predict underperformance in city ecosystems, which I posited as translational ecosystem growth with a reductive management approach.

Over the last three years of UN SDG reporting, Memphis has been consistently ranked in the bottom tier, and in Study 3, the Memphis ecosystem was a case with a pathway to poor sustainable development performance. Memphis' poor performance was predicted with a combination of lack of powerful leadership asset orchestration, meta-organization, university participation, and government support, in addition to having a low percentage of its citizens having climate awareness along with a poor attitude towards sustainability. However, there were examples of sustainable development project success given in my Memphis interviews, for example, an old railway converted into a greenway, a solar panel energy conversion on prison buildings, and the conversion of an old building into a residential mixed-use LEED green certified commercial building. The city as a whole had some degree of change towards sustainability, but efforts did not impact what the SDG reporting measures as key indicators of sustainable cities, which resulted in the poor scores. Moreover, in the Memphis interviews, I consistently heard stories of "the same group of people doing all the project here [Memphis]" (Respondent 13), which indicates fewer partnerships and collaborations were occurring, and fewer opportunities to meta-organize were sensed or seized. What I inferred from these results is that it is not so much the total lack of a meta-organization, for example, but it is the qualitative degree of partnerships, sharing information, drawing from informal local knowledge, and pooling resources fully that illuminates a deficiency in meta-organization. City ecosystems, like my Memphis example, operate in modes of conventional thinking and separated reductive management strategies. There are still aggregated systems and structures of efficiency within the ecosystem, which is equated with translational growth. Referring to (Figure 13) again, I can predict reductive management approaches in city sustainable development that is less agile, more routinized, and aggregated, with top-down only management are less likely to be on a trajectory towards achieving sustainable development goals and sustainable cities. Over time, cities can regress into lower development forms or be stagnate, like in the case of Memphis. I suggest that in order for cities like Memphis, Nashville, Houston, New Orleans, and Oklahoma to improve their sustainable development performance and remedy ecosystem entropy, their primary stakeholders' have to develop a meta-organizing macromanagement approach, like an Appreciative Inquiry Summit, dedicated to achieving developmental goals, inspiring innovation and action, and promoting well-being and flourishing that has the potential to affect the city as a whole.

Whole systems integrative ecosystem macromanagement can yield transformational outcomes and advance higher sustainable development performance for the goal of flourishing and sustainable city ecosystems. Combining diverse reductive and heavily specialized management approaches with informal and indigenous knowledge in local communities appears to be a better approach to managing holistically on the scale of ecosystems. I recommend that a more integrative ecosystem management strategy should be employed by implementing organizations and multi-stakeholders in city ecosystems. City mayors, municipal sustainability offices, academics, citizens, business and community leaders, as well as a diverse mix of stakeholders in the city should consider adopting and participating in forming a meta-organizational governance structure to develop new ecosystem dynamic capabilities and approach sustainable development with a macromanagement approach to:

- promote city-wide flourishing initiatives for the goal of sustainable city
- develop conversations and collaboration to fostering sustainable innovation
- build citizen and organizational participative capacity
- deploy cities' dynamic capabilities through human integration in ecosystems as a function can enable the transformation of various developmental stages with an arching goal towards flourishing CASE

Integrating vertically and horizontally, specialized reductionists knowledge and approaches with indigenous knowledge and approaches should be complementary and form the integrative ecosystem management approach by organizing people, organizations, industries, and institutions within cities and communities for the goal of achieving flourishing CASE.

**Implications**

***Contribution to scholarship.*** I offer this research as a contribution to scholarship by complementing research in ecosystems and community/regional development literature streams, as well as the literature within stakeholder and resilience literature streams. I argue that alternative methods of approaching sustainability and the change expected has not resonated or been adopted on the individual level at a scale necessary for mass flourishing, and that is reflective on the whole image of my cities, not just the good parts. However, individuals believe in the possibilities that they and others have the potential to realize a flourishing potential, which indicates that the AI-SOAR framework can provide the platform and space to improve a community's ability to address, manage, and drive ecological, social, and economic change within the ecosystem.

***Contribution to practice.*** This study to managers in community development, NGOs, innovation, VC/start-up/accelerator practices, as well as municipal institutions, can find this to be useful, as they consider alternative methods of approaching complex systems with the whole ecosystem in mind, instead of approaching problematic issues in the city with siloed and compartmentalized strategies. I do not claim this model as a rigid blueprint to follow, but I do claim that this approach can start the conversation for the reality of change, a conscious shift towards a more balanced societal economy, and promotes the ideation of possibilities of what could be in the future, and of what stakeholders in communities can strive to achieve short term and long-term sustainable development pathways to flourishing. The local business community and private sector can strengthen resource mobilization by forging more partnerships and adopting sustainable practices within their organizations as they integrate resources and systems with deeper collaboration and participation. Universities can also play a critical role as a powerful central innovation ecosystem orchestrating administration. How well universities, business communities, governments, institutions, and all other stakeholders facilitate the sensing and seizing of opportunities, share technical knowledge and other social assets, integrate and coordinate, and accelerate reconfiguring the higher-ordered ecosystem—with operationalizing an AI-SOAR platform—will all moderate the positive contribution towards the transition to the city ecosystem—the CASE.

**Limitations and Future Research**

There are a few limitations in these studies I want to explicate needing to be addressed. In Study 1, the first limitation is that only 4 U.S. cities were sampled from interviews. These cities primarily represented the middle tier to bottom tier rankings of the Sustainable Development Goals Index, which I considered as poor-performing cities relative to their SDG index rankings. By adding a few more sample cities that rank in the higher tiers may potentially add more dynamic insight and gain more lived experiences of successful implementation that may not have been captured from the lower-ranked cities. A second limitation of Study 1 is that this study has focused on the implementation process success rather than on the actual outcomes or impact of a city reaching a sustainable development goal. A third limitation of Study 1 is that managers can subjectively interpret the term success based on experience, and it is difficult to evaluate the subjectivity of what is successful as it relates to sustainable development implementation.

In Study 2, my original model did not result in the separation of the two constructs of Appreciative Inquiry and SOAR as anticipated from previous studies. Instead, they converge into one construct that I labeled AI-SOAR. In addition, I could not achieve full convergent validity of my measurement model, with resilience being the problematic construct; thus, I cautioned further claims in my study for transparency and to maintain the integrity of my work. Further insight could be generated by exploring an individual city or region to understand how community structural changes and innovations emerge from the process uniquely for dedicated city development tasks. For brevity, I did not analyze a moderating effect of my model, and that leaves opportunity for further research to test the moderating effect of AI-SOAR between innovative capacity and resilience. There was also a limitation I must reiterate from the discriminant validity results of the model. It is recommended that AVE factor results should be greater than MVE factor results, and all of my factors did not pass this recommended threshold; however, I do want to mention that other scholars have suggested AVE values are often too strict of a reliability valuation, which also contradicts conventional recommendations. However contradicting recommendations, I still offer caution when considering my claims within that study. Another limitation in my research was that I did not address varying types of innovation, nor speed of designing and stakeholder adoption rate metrics in my study, and a study as such would provide great insight into ecosystem management for inclusive innovation. Additionally, a limitation to this research is the application of

212

the SEM model was generalized, divided, and tested in the variety and diversity of unique cities and their sub-group and subsystem dynamics. A better-designed study can possibly add insight requiring future research to observe my model operated in different ecosystems with varying access to resources and social, environmental, and cultural capitals within the ecosystem. Lastly, an intraclass correlation coefficient (Scartezzini, Germani, & Gritti) analysis may provide a better measurement since my model does assume some nested group properties on the individual level of analysis and emergent effects on the ecosystem level. This is a research plan I tend to explore in future research projects.

There was a limitation in Study 3 I must address. Study 3 only contained samples from U.S. cities only. This may raise concerns about bias and threaten the generalizability of my claims. It also opens an opportunity to test or factor configurations with cities outside of the U.S., especially cities like Finland and Sweden that are globally top ranked. In addition, there may be other varying domain and causal conditions such as the social, political, and economic system structures in place in other cities and countries that may affect results out the U.S. scope. Another limitation was that I used small-N sampling with less than 50 cases in my study. A more robust study would include interviews and secondary data from all 105 evaluated cities in the UN SDG City Report to identify more accurate pathways towards high sustainable development performance.

A limitation of the book is to acknowledge that most of my data collection, analyses, and study completion were performed prior to the COVID-19 pandemic, which has a significant impact on my results and interpretations. The interviews that were conducted in 2017–2018 may have been different as compared to if they were asked post

COVID-19; since the pandemic has significantly impacted economic, social, and environmental systems on a global scale.

Finally, another limitation of the book is that the samples and cases in my studies were all based in the U.S., and responses may have been biased towards a Westernized view of sustainable development, markets, and social system structures. Moreover, there are distinguishable differences in political, social, and economic dynamics in non-Westernized cities, as well as cities in developing countries compared to Westernized cities, which leaves us opportunities to explore research in these areas to observe if my findings are generalizable and relevant to all city ecosystems.

# APPENDIX

## Appendix A: Quotes From Qualitative Study

| | Successful | Unsuccessful | Successful Quote | Unsuccessful Quote |
|---|---|---|---|---|
| **Charlotte** | **Successful** | **Unsuccessful** | **Successful Quote** | **Unsuccessful Quote** |
| GOV | Alternative Energy (Zero emmssions) Conversion in Community | City Center building water waste reduction | Okay... Building a community on the North End ...the entire community residential and commercial would be zero emissions, and totally on an intergrated alternative energy grid. | Absolutely, the water program. We launched the energy program and thought, "Oh, this is how we want to launch every single program. We're going to put up equipment on whatever it is, we're going to measure it, we're going to set a goal and then we're going to put in programs to head towards success, and |
| NGO | Developing City Sustainabilty Plan | Eco. Dev., Health, food security | we've been involved in is the program called Envision Charlotte.... The idea of it was to create a smart grid and within Charlotte began with energy. It has four pillars. It has | We were trying to address food deserts. Although, now I would say that inside of the Seventh Street Public Market our work to address food deserts became difficult because it was hard for us to keep farmers in the Seventh Street Public |
| BIZ | Community bank economic development | Economic Development in depressed area | It has been a while, but the best example I think of success, and this was early on when home ownership is the, was the primary way to build wealth. We created a 100% first time | So we serve that entire area, but we were a special employee group charter, which essentially meant you had to go to each individual employer and get them approved by the regulator that you can serve them. Well, after a number |
| NGO2 | Energy waste reduction | Solid Waste and water waste (individual and | First energy program-program called the Energy Roundtable where we partnered with UNC Charlotte and their students and professors ... They would go into a building and do a | we're looking at our waste and a diversion rate to the landfill. That will have an impact on citizens just in terms of not having a landfill here, the CO2 impact, but also just being a more responsible citizen around waste. But we're not |
| **Chicago** | **Successful** | **Unsuccessful** | **Successful Quote** | **Unsuccessful Quote** |
| GOV | Weatherization program for community (Eco. Dev., Social Equity, stabloe | **Could not think of one** | Rebuilding Together, because it keeps people who have been living in those neighborhoods for 40, 50, 60 years in the neighborhoods another five, six, 10 years. I believe that anything that we can do to help stabilize our neighborhoods | **Could not think of example at time of interview** |
| NGO | Infrastructure conversion to accomodate electric vehicle | Additional amenities for cycling (bike) to promote health | So we're looking at expanding the electric vehicle charging infrastructure on campus, and that come about after we basically had our first commuter transportation survey issued a year ago. And from those findings, we learned that | So we have a large cycling community, that they've requested additional amenities. More showers, more indoor parking, and so forth. We have to partner up with multiple stakeholders on campus even just considering what those amenities look like. So our transportation facilitator, one of my |
| ACA | Facilities Mgmt Energy conversion to zero emmision | **Alternative energy conversion and pollution** | ...we'd be a fully wealthy school, and we've got a goal to be net-zero by the year 2050.. For (school), we're very research intensive. Research is really our forte, and research intensity means that we're a little bit more energy intense than a | I'll never have enough space to have enough renewable generation to offset all of my energy usage. We're space poor. And same thing with (school), and most universities will never have enough land to be able to have that much |
| NGO2 | Buildings energy waste reduction | University healthcare campus facilities energy and waster reduction | The project had some obstacles logistics wise in the beginning and they had to figure out how to store and pick up recyclables, but once an alliance was built with the county workforce development NGO, a solution to add green jobs to people whom were out of work was developed, and it reduced waste in the health system, and created a circular economy with the use and reuse of plastics, and reduced | This leader did not believe in climate change from a religious perspective, and when you don't have buy-in from the leadership above, it makes it hard to initiate or get pull through for projects identified to bring sustainable solutions to the organization. She also mentioned the lack of infrastructure on the sustainability movement and Government side |
| **Cleveland** | **Successful** | **Unsuccessful** | **Successful Quote** | **Unsuccessful Quote** |
| GOV | Affordable housing, weatherization (energy waste reduction) | Community Engagement for Sustainable projects | Our office, we work in two big buckets. One is internal within city operations, so how we lead by example and sustainability within all our departments, right? ... That's public works, that's public utilities, the airport, the streets, | Oh yeah. I think with any sustainability program, the challenge is getting beyond acquire. Our acquire inquiry is larger than most cities, but it's still ... If you ask the average person in Cleveland do they know what Sustainable Cleveland is, they probably wouldn't know that. I think they would probably be |
| NGO | ...land use convert to urban farm (Eco. Dev. Work Force Dev., | Land Use decision making conflict city | ...it's really easy to talk about just in the amazing energy that was created when we first launched our reimagining more sustainable Cleveland work. It was the culmination of about a year long collaborative, with every | So two examples, aside from the ones I already mentioned which was that over the last seven years, the enthusiasm for market gardening has tapered. The two I would share with you is we started in early 2014, we started a |
| BIZ | Economic development program with creative financing (Eco. Dev., | Economic Development initiative with high risk | trying to drive increased involvement of our employees internally with the community. So utilizing our existing business relationships and tying, trying to tie those relationships to non-profits locally. So for example, we | So ultimately, I mean yes, the community bank and the corporate bank are both, we're a for-profit, we're a bank. So ultimately yeah, we do have to deliver to stakeholders and, you know, make money. But in the long run, yeah. I think it would definitely be. I mean, we can't say that it'll definitely be beneficial, but |
| BIZ2 | Organization waste management program recycle program | Engaging (transparency) employee project | So in the spirit of being responsible as an organization, we have ... obviously we have recycling containers and things like that around the building. Some get utilized effectively, some don't. So my IT manager and I put together a program | So we ... actually coming from that conference we were at, we had kind of walked away with plans of creating an internal communication forum in the shape of a podcast, which was to be run by our employees, for our employees. So giving them a voice to communicate issues, learn things from each other, |
| **Memphis** | **Successful** | **Unsuccessful** | **Successful Quote** | **Unsuccessful Quote** |
| GOV | Correctional facility Building Mgmt- Alternative energy conversion | Engaging community on new sustainability projects, & Not | I will mention would be the Mid-South Regional Greenprint, which although it's not development, it's about developing a network of greenways and trails, and there was a planning initiative where we coordinated in a four county area, three states - Arkansas, Mississippi, and then Tennessee, 18 | Yeah. Yeah, there were a few. There was this group of, I guess they call themselves Tea Partiers, so it's not me labeling them. They're part of the Shelby County Tea Party, and they had some reservations about any type of sustainable activity. They believe that Agenda 21, which was something that was passed by the UN ... It was one of these resolutions that they passed in the |
| NGO | Greenways and parks- Land Use | Alternative energy conversion and pollution | The whole premise behind that project is that as a non-profit organization, essentially, is to roll back into the programs whether they be around sustainability, health, and all that kind of falls under the same thing. Any money that's made | Sierra Club national actually got some huge contributions that they decided they wanted to focus energy on reducing coal-based energy, and so, they've actually hired people. The name of the campaign is Beyond Coal, which partly means what other energy sources but it also means let's get rid of coal. At one |
| BIZ | Sustainable Product Service Innovation (Eco. Dev.) | Engaging workforce to participate in sustainable | primarily it's the Innovation Program that IT has started two years ago.....They weren't coming to them with new ideas. (protect and alleging 00:03:30), because the business is constantly asking IT for new ideas. We were so focused on | I think part of it is that they're all part-time. They don't do a whole lot in innovation most of the month and then they attend our, we call it a Plank Meeting. Innovation is one of the five strategic planks for IT. We talk about the program, how it's going. I present metrics. I talk about the ideas that are |
| ACA | Obtaining Star Community Rating (establishing metric system for | City buildings (fixed assests) energy waste reduction | I would say the Star Community that I mentioned earlier as a membership is also when we submitted for that rating system we received a four star rating which is there's a five star that I think is the highest you can get to. It's an ongoing | it's not really an initiative but I think just in terms of it's been a goal in a lot of our planning processes over the last six years to look at government leading by example in terms of our building energy use and efficiency. I feel like there has been some success stories on that front, but I feel like overall there hasn't been |
| NGO2 | Energy projects | Community Engagement for Sustainable projects | We have a lot of projects actually out there. I wouldn't be able to go through all of them. I wouldn't be able to remember all of them, but you have the agri-center, speaking of agriculture. It has installed a large solar array, I | I think this needs to be approached again with lessons learned. I think that last year we did a campaign called We Sustain Shelby which was about getting the community involved and actions that they could do at home or within their daily lives that were sustainable. Whether it be changing their water out to |
| **ADHOC** | **Successful** | **Unsuccessful** | **Successful Quotes** | **Unsuccessful Quote** |
| GOV | Economic Development building transportation and energy system for community (job | Eco. Dev. Project with tension between gov. officals and contractor | These are small, extremely rural communities, not connected by the road system. What is now the Northwest Arctic Borough, didn't exist when this project began. It didn't have any revenues and therefore was an unorganized borough. It didn't have real responsibilities. This project created 500 jobs | I don't know if that's a failure, but I think there's a lesson learned in that in terms of what could we do differently. One, a fully staffed office would help. But, two, I think marketing around that and just helping to ... I think you can't just run something like that once. Sometimes you've got to address it and now maybe we might focus on a different target group like corporations because |
| VOL | Rebuilding affordable homes in community after catastrophe | Leadership Choosing Communities to Rebuild after | I was a volunteer in New Orleans for about a week. It was a couple years after Katrina. I went down with Deutsche Bank. I took my own vacation time, and I paid out of pocket to do this group volunteer work. So, basically signed up to | When everything was said and done, they were getting help. It just seemed like certain communities were still being neglected. I know that when I went down there, I met with a photographer who lost most of his work, but he was working on ... He knew Brad Pitt personally, and he basically mentioned that |

215

# Appendix B: Qualitative Sample City/Interviewee Demographics

| Participant Demographics | | | | | | | |
|---|---|---|---|---|---|---|---|
| **Charlotte** | Gender | Race | Education Level | Education Background | Age range | Position | Industry |
| GOV | M | Caucasian | Masters | Engineering | 35-45 | Director/Officer | Municipal Agent |
| NGO | W | Caucasian | College | Arts and English | 45-55 | Executive Director | Local Non Profit |
| BIZ | M | Caucasian | College | Finance and Accounting | 45-55 | Vice President | Community Bank |
| NGO2 | W | Caucasian | College | Communication | 45-55 | Manager | Local Non Profit |
| **Chicago** | | | | | | | |
| GOV | M | African American | Masters | MBA | 35-45 | Director | Municipal Agent |
| NGO | W | Hispanic American | Masters | Achitecture and Sustainablity | 35-45 | Manager | Local Non Profit |
| ACA | M | Caucasian | Doctorate | Engineering | 45-55 | Vice President | Academic Institution |
| NGO2 | W | Caucasian | Masters | Healthcare Management | 35-45 | Manager | Academic Institution |
| **Cleveland** | | | | | | | |
| GOV | M | Caucasian | Masters | Engineering | 35-45 | Director/Officer | Municipal Agent |
| NGO | W | Caucasian | Masters | MBA | 45-55 | Executive Director | Local Non Profit |
| BIZ | W | Asian American | College | Business | 25-35 | Manager | Large National Bank |
| BIZ2 | M | Caucasian | College | Accounting | 25-35 | Executive Manager | Small/Medium Media |
| **Memphis** | | | | | | | |
| GOV | M | African American | Masters | Engineering, Real Estate, Finance | 35-45 | Director | Municipal Agent |
| NGO | M | Caucasian | Masters | MBA/Planning | 55-65 | District Chairperson | National Non-Profit |
| BIZ | M | Caucasian | Masters | IT | 45-55 | Manager | Multinational Corporation |
| ACA | W | Caucasian | Doctorate | Architecture | 35-45 | Scholar/ Director | Municipal Agent/Scholar |
| NGO2 | W | Caucasian | Masters | Planning | 35-45 | Manager | Local Non Profit/Municipal Agent |
| **ADHOC** | | | | | | | |
| GOV | M | Caucasian | Doctorate | Engineering | 45-55 | Executive Director | Economic Development Corp |
| VOL | W | Hispanic American | Masters | MBA | 45-55 | Manager | International Bank |

| Gender | Race | Education Level | Education Background | Age range | Position | Industry |
|---|---|---|---|---|---|---|
| Men = 49% | 76% Caucasian | 100% College Degree | 30% - Engineering | 10% - (25-35) | 43% - Management | **Business Size** |
| Women = 51% | 10 % African American | 24 % Masters | 30% - MBA, Business or Health Management | 52% - (35-45) | 19 % -Executive Director | 50% - Small/Medium |
| | 10% HispanicAmerican | 15% Doctorate Degree | 10% - Architecture | 29% - (45-55) | 19 % Director | 50% - Large/MNC |
| | 5% Asian American | | 10% - Urban Planning | 9 % - (55-65) | 14% - Vice President | **Business Types** |
| | | | 10% - Finance/Accounting | | 5% -Chairperson (Regional) | 90% - Banking |
| | | | 10% - Libral Arts | | | 5% - Manufacturing |
| | | | | | | 5% -Media Services |

| Charlotte | Chicago | Cleveland | Memphis | Ad hoc |
|---|---|---|---|---|
| Participant | Participant | Participant | Participant | Participant |
| GOV | GOV | GOV | GOV | GOV |
| NGO | NGO | NGO | NGO | NGO |
| BIZ | BIZ | BIZ | BIZ | BIZ |
| ACA | ACA | ACA | ACA | ACA |
| NGO2 | BIZ2 | BIZ2 | NGO2 | GOV |
| VOL | VOL | VOL | VOL | VOL |

**Codes***: Gov-* Government, *NGO-* Non-profit/governmental organization, *BIZ-* Business firm, *ACA-* Academic, *VOL-* Volunteer/Community Activists

# Appendix D: Qualitative Thematic Coding Process

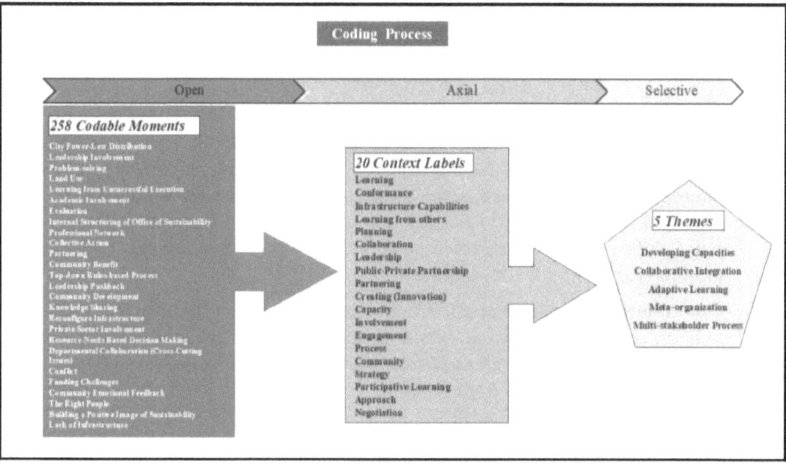

**Appendix E: Qualitative Interview Protocol and Interview Questions**

**Introduction (interviewer):** *"Hello my name is Larry Clay. I am a doctoral student at Case Western Reserve University. Thank you so much for taking the time to meet with me today. I appreciate it. Before getting started, there are a couple of things I would like to cover."*

**Purpose and Format for the Interview (Interviewer):** *"As a current student in the Case Western Reserve University Doctorate of Management (DM) program, I am interested in developing a greater understanding of how sustainability is developed and implemented within cities. I will ask you a series of open-ended questions on this topic, and I will also ask one or more follow-up questions as you respond. The interview will last approximately 60 – 90 minutes."*

**Confidentiality (Interviewer):** *"Everything you share in this interview will be kept in strictest confidence, and your comments will be transcribed anonymously – omitting our names, anyone else you refer to in this interview, as well as the name of your current organization and/or past organizations. Your interview responses will be included with all the other interviews I conduct."*

**Audio Taping (Interviewer):** *"To help me capture your responses accurately and without being overly distracting by taking notes, I would like to record our conversation with your permission. Again, your responses will be kept confidential. If at any time, you are uncomfortable with this interview, please let me know and I will turn the recorder off."*

*"Any questions before we begin?*

1. **First, I'd like to learn a little about you. Would you give me a brief bio about yourself?**

   Probes:
   - Education, job history, family, interests, where live, international travel.
   - Do you volunteer for anything?
   - Please tell me about your current role.
   - What is the business, the organization, your position, how long have you been with the company/office, what has been your most accomplished work, thus far?
   - Where do the local city sustainable development goals come from and what do you use as a metric?
   - Were they directed as national, state or local level initiatives?
   - How many initiatives are currently in Implementing- Monitoring phases?
   - What is the largest initiative conducted to date?

- Who Plans? Coordinates? Implements? Provides resources?
- What did you learn from this experience?

**2. Can you give me an example when you were involved in a sustainable development initiative with outcomes that were successful?**

<u>Probes:</u>
- Who else was involved (internal and external)?
- How was the community benefitted?
- What was the positive outcome?
- What was the inspiration?
- When did it occur?
- How long did it last?
- Describe some of the barriers or challenges you have faced when implementing this initiative.
- Were there any stakeholders against the initiative?
- Anything that would have made it better?
- What made this a success?
- What was the most impactful sustainable development implemented in the city to date?
- What did you learn from this experience?

**3. Can you provide an example when you were involved in a sustainable development initiative with outcomes that were less than successful?**

<u>Probes:</u>
- Who else was involved (internal and external)?
- How was the community benefitted?
- What was the positive outcome?
- What was the inspiration?
- When did it occur?
- How long did it last?
- Describe some of the barriers or challenges you have faced when implementing this initiative.
- Were there any stakeholders against the initiative?
- Anything that would have made it better?
- What did you learn from this experience?

4. **Could you provide an example of a metro area sustainable development program that is planned and on the verge of implementation?**

   <u>Probes:</u>
   - Description
   - Describe the initial challenges
   - Who are the committed stakeholders?
   - Who are the not so committed stakeholders?
   - Goals of this program
   - Metrics
   - National, State or local scope
   - What will be your role in the process? Planning, Strategy, Implementation, Partnership Development, or Monitoring/Evaluation?

**Conclusion**
"Thank you so much for our conversation! I've had an enjoyable time and I hope you have as well. This has been extremely helpful. One last thing - Is there anything else that we haven't discussed that you would like to add before we finish?
We have covered a lot of ground, but if by any chance I have missed anything, would it be OK to contact you again for a brief clarification on any particular points, if necessary?"
"Thank you for your time today. Can you think of anyone else I should contact for an interview? Please know that you are under no obligation to provide this information."

## Appendix F: Quantitative Survey Instrument

### <u>Innovative Capacity (Reflective)</u>

**Definition:** The managing of creativity and innovation outcomes through stakeholders' cognitive energy, reward, recognition, and objective satisfaction

**Instructions:** On a scale of 1-5, with 5 being the highest score: compared to other cities, how would you perceive you and other stakeholders in your CASE successful pursuit of the following performance indicators:

Stakeholders in our ecosystem suggest new ways to achieve goals or objectives
Stakeholders in our ecosystem come up with new and practical ideas to improve performance
Stakeholders in our ecosystem are good sources of creative ideas
Stakeholders in our ecosystem are not afraid to take risks
Stakeholders in our ecosystem search outside our network for new technologies, processes, techniques, and/or product ideas*
Stakeholders in our ecosystem suggest new ways to increase quality of life
Stakeholders in our ecosystem often have fresh approaches to issues
Stakeholders in our ecosystem develop adequate plans and schedules for the implementation of new ideas*
Stakeholders in our ecosystem promote and champions ideas to others*
Stakeholders in our ecosystem often have new and innovative ideas
Stakeholders in our ecosystem come up with creative solutions
Stakeholders in our ecosystem suggest new ways of performing task
Stakeholders in our ecosystem exhibit creativity when given the opportunity
Stakeholders in our ecosystem are good sources of creative ideas
*Indicates measurement items developed from Scott & Bruce, 1994*
- Scale developed by George & Zhou, 2002

### <u>Resilience</u>

**Definition:** 1) A positive psychological system capacity to rebound, to 'bounce back' from adversity, uncertainty, conflict, failure, or other perturbation. 2) The capacity of a system to undergo change and still retain its basic function and structure.

**Instructions:** On a scale of 1-5, with 5 being the highest score- Compared to other cities, how would you perceive your city's successful pursuit of the following performance indicators:

Can access funds for dealing with short-term disasters
Can access insurance coverage for major public and private assets
Has a diverse economy and workforce
Has opportunities for education, training and learning
Has leaders who adjust quickly to change
Has strong leaders who work well together

Is made up of people who support each other
Is made up of people who trust each other
Has long-term plans aimed at ensuring a diversified local economy
Has long-term plans that aim to manage natural resources sector development
Has long-term plans that aim to manage sustainable development
Integrates and shares knowledge amongst stakeholders
Is regularly informed about changes affecting the City as an Ecosystem
Participates in risk and vulnerability planning
Plans for disasters, loss, hazards, vulnerabilities and risk
Prepares and trains for long-term change
Prepares and trains for short-term change
Works well together across internal and external bodies
*      Scale developed by Bec, 2018

### <u>Appreciative Inquiry [AI] (the operating system) &<br><u>Strengths, Opportunities, Aspire, & Results [SOAR] (the process)</u><br><u>AI-SOAR (Reflective)</u>

**Definition:** The operating system and strategic process of a macro-management, whole systems  strengths-based approach that focuses on strengths and opportunities instead of weaknesses and problematic issues for system-level solutions.

**Instructions:** For each of the following 19 items, select one of the response options from "never" to "always": ○ Never ○ 20% of the time ○ 30% of the time ○ 40% of the time ○ 50% of the time ○ 60% of the time ○ 70% of the time ○ 80% of the time
○ 90% of the time ○ Always
When you approach strategy in your life and City as an Ecosystem, how often do you focus on Collaborative Relationships? "Collaborative Relationships" involve working together in your City as an Ecosystem to achieve shared goals.

When you approach strategy in your life and City as an Ecosystem, how often do you focus on Generative Questions? "Generative Questions" are questions you ask to discover or create new things that you can use to positively alter the collective future of yourself or your City as an Ecosystem.

When you approach strategy in your life and City as an Ecosystem, how often do you focus on Open Communications? "Open Communication" is the straightforward and truthful communication by all parties in your City as an Ecosystem to express ideas to each another.

When you approach strategy in your life and City as an Ecosystem, how often do you focus on Positive Framing? "Positive Framing" refers to a positive frame of thought that focuses on overcoming challenges, motivating people, and moving collaborative projects forward for you or your City as an Ecosystem.

When you approach strategy in your City as an Ecosystem, how often do you focus on Solutions? "Solutions" are creating opportunities to achieve success for you or your City as an Ecosystem.

When you approach strategy in your City as an Ecosystem, how often do you focus on Stakeholder Needs? "Stakeholders' Needs" are the interests or needs of those people internal or external to the City as an Ecosystem essential to achieving the communities' success.

When you approach strategy in your City as an Ecosystem, how often do you focus on Whole systems? "Whole systems " is how things are related and how they influence one another within your City as an Ecosystem.

When you approach strategy in your life and City as an Ecosystem, how often do you focus on Assets? "Assets" are those strengths that create personal and total City as an Ecosystem value.

When you approach strategy in your life and City as an Ecosystem, how often do you focus on Capabilities? "Capabilities" are those abilities that create the best for yourself and your City as an Ecosystem.

When you approach strategy in your life and City as an Ecosystem, how often do you focus on Strengths? "Strengths" are those greatest capabilities of you and your City as an Ecosystem.

When you approach strategy in your life and City as an Ecosystem, how often do you focus on Ideas? "Ideas" are your thoughts or suggestions for possible courses of action for yourself and your City as an Ecosystem.

When you approach strategy in your life and City as an Ecosystem, how often do you focus on Opportunities? "Opportunities" are those ideas or innovations that make it possible to turn personal and whole City as an Ecosystem visions into reality.

When you approach strategy in your life and City as an Ecosystem, how often do you focus on Possibilities? "Possibilities" are those outcomes that may lead to personal and whole City as an Ecosystem success.
When you approach strategy in your City as an Ecosystem, how often do you focus on Aspirations? Aspirations are your strong desires to achieve the whole City as an Ecosystem vision.

When you approach strategy in your life and City as an Ecosystem, how often do you focus on Desires? "Desires" are your sense of hope for success for yourself and your whole City as an Ecosystem.

When you approach strategy in your life and City as an Ecosystem, how often do you focus on Values? "Values" are what you and your whole City as an Ecosystem care deeply about.

When you approach strategy in your City as an Ecosystem, how often do you focus on Completed Tasks? "Completed Tasks" are those completed activities that help you achieve results for your whole City as an Ecosystem.

When you approach strategy in your City as an Ecosystem, how often do you focus on Outcomes? "Outcomes" are those meaningful activities that when completed will have a positive result for yourself and your whole City as an Ecosystem.

When you approach strategy in your City as an Ecosystem, how often do you focus on Results? "Results" are the measured total outcomes for success.

- Scale developed by Stavros & Cole, 2018

# Appendix G: Sample Demographics of Quantitative Study

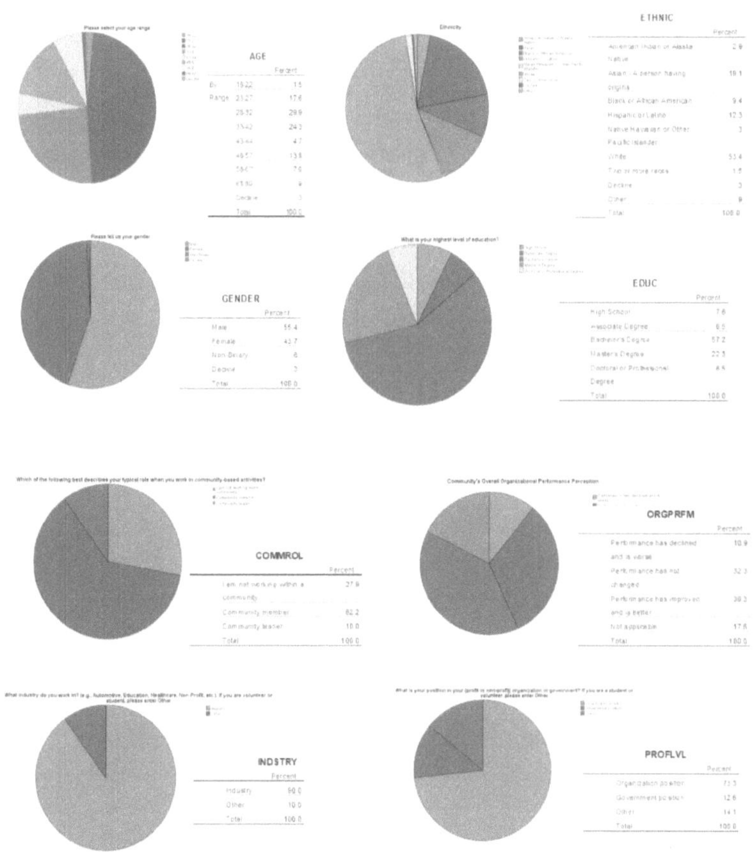

| Total Variance Explained | | |
|---|---|---|
| Factor | | |
| Total | Cumulative % | Total% |
| 1 | 46.446 | 46.446 |
| 2 | 5.08 | 51.526 |
| 3 | 3.004 | 54.53 |

**Appendix I: Quantitative Unobserved Variable Correlation Matrix**

| Factor | AI-SOAR | Innovative Capacity | Resilience |
|---|---|---|---|
| **AI-SOAR** | - | | |
| **Innovative Capacity** | 0.680 | - | |
| **Resilience** | 0.732 | 0.744 | - |

# Appendix J: Quantitative Three-Factor Pattern Matrix

| | 3 Factor | | |
|---|---|---|---|
| | AI-SOAR | Innovative Capacity | Resilience |
| RRA1 | | | 0.404 |
| RRA2 | | | 0.373 |
| RDI1 | | | 0.545 |
| RDI2 | | | 0.408 |
| RCT1 | | | 0.524 |
| RCT2 | | | 0.722 |
| RCT3 | | | 0.486 |
| RCT4 | | | 0.568 |
| RCC1 | | | 0.656 |
| RCC2 | | | 0.688 |
| RCC3 | | | 0.736 |
| RCC4 | | | 0.537 |
| RCC5 | | | 0.613 |
| RCC6 | | | 0.581 |
| RCC7 | | | 0.589 |
| RCC8 | | | 0.716 |
| RCC9 | | | 0.621 |
| RCC10 | | | 0.772 |
| IC1 | | 0.589 | |
| IC2 | | 0.642 | |
| IC3 | | 0.804 | |
| IC4 | | 0.743 | |
| IC5 | | 0.607 | |
| IC6 | | 0.701 | |
| IC7 | | 0.77 | |
| IC8 | | 0.633 | |
| IC9 | | 0.59 | |
| IC10 | | 0.769 | |
| IC11 | | 0.772 | |
| IC12 | | 0.703 | |
| IC13 | | 0.75 | |
| IC14 | | 0.86 | |
| AI1 | 0.767 | | |
| AI2 | 0.697 | | |
| AI3 | 0.786 | | |
| AI4 | 0.681 | | |
| AI5 | 0.71 | | |
| AI6 | 0.79 | | |
| AI7 | 0.611 | | |
| SST1 | 0.692 | | |
| SST2 | 0.586 | | |
| SST3 | 0.879 | | |
| SOP1 | 0.736 | | |
| SOP2 | 0.69 | | |
| SOP3 | 0.689 | | |
| SAS1 | 0.786 | | |
| SAS2 | 0.7 | | |
| SAS3 | 0.702 | | |
| SRE1 | 0.666 | | |
| SRE2 | 0.636 | | |
| SRE3 | 0.652 | | |
| Cronbach α | 0.962 | 0.904 | 0.935 |

**Appendix K: fsQCA City Case Characteristics and Domain Conditions**

| CaseID | Rank section 2019 | Index Score2017 | Index Score 2018 | Index score 2019 | 3-Yr. Index Δ | R1-R2 Univ. | MSA Pop (mil) | GMP ($mil) | MGDP Grwth | Patents | Local Adpt Plan | Climate Aware | %B_lo_Pvrty Line | Wrkng Poor | Upwrd Mbl | Racial Seg | STEM employment | Sustain_transit. |
|---|---|---|---|---|---|---|---|---|---|---|---|---|---|---|---|---|---|---|
| San Francisco, CA | 1 | 56.43 | 61.7 | 69.7 | 13.27 | 2 | 4.7 | 548.6 | 83 | 100 | 100 | 100 | 70 | 80 | 80 | 47 | 91 | 66 |
| Washington DC | 1 | 50.31 | 58.44 | 66.7 | 16.39 | 8 | 6.2 | 540.6 | 23 | 16 | 100 | 87 | 76 | 81 | 67 | 79 | 93 | 45 |
| Austin, TX | 1 | 52.57 | 51.16 | 61.8 | 9.23 | 1 | 2.2 | 146.8 | 100 | 75 | 100 | 74 | 61 | 67 | 37 | 82 | 71 | 12 |
| Denver, CO | 2 | 49.85 | 50.44 | 54.5 | 4.65 | 2 | 3 | 214.1 | 60 | 23 | 100 | 64 | 72 | 73 | 54 | 58 | 67 | 16 |
| Salt Lake City, UT | 2 | 48.75 | 43.55 | 53.8 | 5.05 | 2 | 1.2 | 94.3 | 45 | 39 | 100 | 39 | 70 | 71 | 92 | 38 | 52 | 21 |
| Chicago, IL | 2 | 40.16 | 43.84 | 52.6 | 12.44 | 7 | 9.5 | 660 | 27 | 30 | 100 | 77 | 53 | 70 | 33 | 10 | 40 | 37 |
| Pittsburg, PA | 3 | 47.2 | 50.85 | 50.5 | 3.3 | 3 | 2.3 | 152.8 | 46 | 21 | 100 | 31 | 58 | 77 | 83 | 22 | 70 | 20 |
| Philadelphia, PA | 3 | 39.49 | 45.96 | 49.7 | 10.21 | 5 | 6.1 | 444.2 | 36 | 23 | 100 | 66 | 48 | 73 | 44 | 37 | 68 | 29 |
| Charlotte, NC | 3 | 42.82 | 43.8 | 49.5 | 6.68 | 1 | 2.6 | 169.9 | 57 | 11 | 0 | 36 | 52 | 64 | 0 | 71 | 33 | 6 |
| Kansas City, MO | 4 | 43.28 | 43.48 | 49 | 5.72 | 1 | 2.1 | 132.7 | 29 | 22 | 0 | 44 | 64 | 71 | 39 | 65 | 63 | 1 |
| Atlanta, GA | 4 | 43.07 | 46.89 | 48.7 | 5.63 | 4 | 6 | 397.3 | 66 | 26 | 0 | 47 | 44 | 59 | 0 | 70 | 39 | 13 |
| Nashville, TN | 4 | 43.61 | 48.8 | 47.7 | 4.09 | 2 | 1.9 | 132.2 | 81 | 5 | 0 | 16 | 58 | 71 | 20 | 87 | 36 | 6 |
| Detroit, MI | 5 | 31.82 | 43.12 | 46.5 | 14.68 | 4 | 4.3 | 267.7 | 35 | 52 | 100 | 40 | 37 | 61 | 11 | 8 | 67 | 1 |
| Cleveland, OH | 5 | 31.41 | 37.23 | 45.4 | 13.99 | 3 | 2.1 | 134.3 | 39 | 30 | 100 | 52 | 36 | 66 | 17 | 12 | 56 | 4 |
| Jacksonville, FL | 5 | 37.06 | 42.67 | 45.1 | 8.04 | 0 | 1.6 | 83.2 | 61 | 9 | 100 | 26 | 45 | 63 | 13 | 71 | 46 | 6 |
| St. Louis, MO | 6 | 38.9 | 41.61 | 42.9 | 4 | 3 | 1.8 | 169.8 | 36 | 15 | 0 | 40 | 54 | 70 | 28 | 32 | 54 | 4 |
| Houston, TX | 6 | 40.79 | 38.12 | 41.9 | 1.11 | 3 | 7 | 478.4 | 27 | 33 | 100 | 42 | 41 | 46 | 65 | 64 | 52 | 11 |
| Las Vegas NV | 6 | 45.85 | 42.53 | 40.4 | -5.45 | 1 | 2.7 | 122.4 | 53 | 24 | 0 | 63 | 47 | 55 | 36 | 100 | 0 | 12 |
| Oklahoma City OK | 7 | 44.58 | 42.48 | 40.3 | -4.28 | 1 | 1.4 | 81 | 53 | 6 | 0 | 21 | 41 | 54 | 39 | 76 | 47 | 7 |
| New Orleans, LA | 7 | 33.16 | 30.49 | 39.8 | 6.64 | 1 | 1.3 | 80.787 | 32 | 5 | 100 | 42 | 34 | 39 | 20 | 43 | 36 | 13 |
| Memphis, TN | 7 | 34.39 | 39.11 | 35.8 | 1.41 | 1 | 1.4 | 76.8 | 6 | 18 | 100 | 34 | 20 | 31 | 0 | 40 | 36 | 1 |

**Appendix L: Direct Calibration Condition**

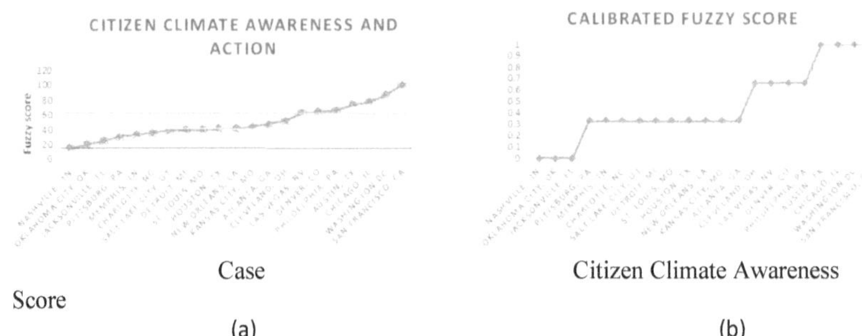


(a)

(b)

**Citizen climate, culture, and awareness* scores, which reflect the estimated percentage of adults who think global warming is happening and human intervention is needed to address it, was calculated using the UN SDG indicator 13.3- Climate awareness and action (Lynch, SDG 17 codebook, 2020) and sourced from the Yale Climate Opinion Maps, Yale Program on Climate Change Communication. In-set membership is when the total case CCCAA score is greater than 0.7 (green line). Out-of-set membership is when the total case CCCAA score is less than 0.2 (red). The crossover point is when the total case CCCAA score is 0.5 (yellow line), which reflects a break between the case scores.

(a) Anchor and crossover points for direct calibration: *Citizen climate, culture, and awareness.*
(b) Direct Calibration for *Citizen climate, culture, and awareness.*

# Appendix M: Outcome Indirect Calibration

**Indirect Calibration of Total City
Sustainable Development Performance Scores**

| Outcome | Calibration | |
|---|---|---|
| *Total City Performance Score change over 3yr* | 0 | City total 3 yr. score change was a unit 1 points or less (stagnation) or negative (declining) values |
| | 0.33 | City total 3 yr. score unit change was greater than 2 but less than 6 points |
| | 0.66 | City total 3yr score unit change was greater than 6 but less than 10 points |
| | 1 | City total 3 yr. score improved by 10 or greater unit points |

Data extracted from the Annual UN SDG Sustainable City Achievement (Lynch, 2019)

# REFERENCES

Adler, A. & Seligman, M. E. 2016. Using wellbeing for public policy: Theory, measurement, and recommendations. *International Journal of Wellbeing*, 6(1): 1-35.

Adner, R. & Helfat, C. E. 2003. Corporate effects and dynamic managerial capabilities. *Strategic Management Journal*, 24(10): 1011-1025.

Adner, R. & Kapoor, R. 2010. Value creation in innovation ecosystems: How the structure of technological interdependence affects firm performance in new technology generations. *Strategic Management Journal*, 31(3): 306-333.

Adner, R. 2017. Ecosystem as structure: An actionable construct for strategy. *Journal of Management*, 43(1): 39-58.

Ahrne, G. & Brunsson, N. 2008. *Meta-organizations*. Cheltenham, UK: Edward Elgar Publishing.

Albareda-Tiana, S., Vidal-Raméntol, S., & Fernández-Morilla, M. 2018. Implementing the sustainable development goals at University level. *International Journal of Sustainability in Higher Education*, 19(3): 473-497.

Almirall, E., Lee, M., & Majchrzak, A. 2014. Open innovation requires integrated competition-community ecosystems: Lessons learned from civic open innovation. *Business Horizons*, 57(3): 391-400.

Anderson, J. C. & Gerbing, D. W. 1988. Structural equation modeling in practice: A review and recommended two-step approach. *Psychological Bulletin*, 103(3): 411-423.

Angelidou, M. 2014. Smart city policies: A spatial approach. *Cities*, 41: S3-S11.

Appio, F. P., Lima, M., & Paroutis, S. 2019. Understanding Smart Cities: Innovation ecosystems, technological advancements, and societal challenges. *Technological Forecasting and Social Change*, 142: 1-14.

Armitage, D., Berkes, F., & Doubleday, N. (Eds.). 2010. *Adaptive co-management: Collaboration, learning, and multi-level governance*. Vancouver, BC, Canada: UBC Press.

Aryee, S., Walumbwa, F. O., Zhou, Q., & Hartnell, C. A. 2012. Transformational leadership, innovative behavior, and task performance: Test of mediation and moderation processes. *Human Performance*, 25(1): 1-25.

Bäckstrand, K. 2006. Multi-stakeholder partnerships for sustainable development: Rethinking legitimacy, accountability and effectiveness. *European Environment*, 16(5): 290-306.

Bagozzi, R. P. & Yi, Y. 2012. Specification, evaluation, and interpretation of structural equation models. *Journal of the Academy of Marketing Science*, 40(1): 8-34.

Bai, X., Wieczorek, A. J., Kaneko, S., Lisson, S., & Contreras, A. 2009. Enabling sustainability transitions in Asia: The importance of vertical and horizontal linkages. *Technological Forecasting and Social Change*, 76(2): 255-266.

Bai, X., Surveyer, A., Elmqvist, T., Gatzweiler, F. W., Güneralp, B., Parnell, S., Prieur-Richard, A.-H., Shrivastava, P., Siri, J. G., & Stafford-Smith, M. 2016. Defining and advancing a systems approach for sustainable cities. *Current Opinion in Environmental Sustainability*, 23: 69-78.

Bain, P. G., Kroonenberg, P. M., Johansson, L.-O., Milfont, T. L., Crimston, C. R., Kurz, T., Bushina, E., Calligaro, C., Demarque, C., & Guan, Y. 2019. Public views of the sustainable development goals across countries. *Nature Sustainability*, 2(9): 819-825.

Basurto, X. & Speer, J. 2012. Structuring the calibration of qualitative data as sets for qualitative comparative analysis (QCA). *Field Methods*, 24(2): 155-174.

Bayulken, B. & Huisingh, D. 2015. A literature review of historical trends and emerging theoretical approaches for developing sustainable cities (part 1). *Journal of Cleaner Production*, 109: 11-24.

Bec, A., Char-lee, J. M., & Moyle, B. D. 2019. Community resilience to change: Development of an index. *Social Indicators Research*, 142(3): 1103-1128.

Becht, G. 1974. Systems theory, the key to holism and reductionism. *Bioscience*, 24(10): 569-579.

Bergandi, D. & Blandin, P. 1998. Holism vs. reductionism: Do ecosystem ecology and landscape ecology clarify the debate? *Acta Biotheoretica*, 46(3): 185-206.

Berke, P., Backhurst, M., Day, M., Ericksen, N., Laurian, L., Crawford, J., & Dixon, J. 2006. What makes plan implementation successful? An evaluation of local plans and implementation practices in New Zealand. *Environment and Planning B: Planning and Design*, 33(4): 581-600.

Berke, P. R. & Conroy, M. M. 2000. Are we planning for sustainable development? An evaluation of 30 comprehensive plans. *Journal of the American Planning Association*, 66(1): 21-33.

Berkowitz, H. & Dumez, H. 2016. The concept of meta-organization: Issues for management studies. *European Management Review*, 13(2): 149-156.

Berkowitz, H., Bucheli, M., & Dumez, H. 2017. Collectively designing CSR through meta-organizations: A case study of the oil and gas industry. *Journal of Business Ethics*, 143(4): 753-769.

Berkowitz, H. 2018. Meta-organizing firms' capabilities for sustainable innovation: A conceptual framework. *Journal of Cleaner Production*, 175: 420-430.

Berkowitz, H. & Bor, S. 2018. Why meta-organizations matter: A response to Lawton et al. and Spillman. *Journal of Management Inquiry*, 27(2): 204-211.

Berkowitz, H. & Souchaud, A. 2019. (self-)regulation of sharing economy platforms through partial meta-organizing. *Journal of Business Ethics*, 159(4): 961-976.

Berkowitz, H., Crowder, L. B., & Brooks, C. M. 2020. Organizational perspectives on sustainable ocean governance: A multi-stakeholder, meta-organization model of collective action. *Marine Policy*, 118: 10.

Berman, E. & Wang, X. 2000. Performance measurement in US counties: Capacity for reform. *Public Administration Review*, 60(5): 409-420.

Bingham, L. B., Nabatchi, T., & O'Leary, R. 2005. The new governance: Practices and processes for stakeholder and citizen participation in the work of government. *Public Administration Review*, 65(5): 547-558.

Bogers, M., Chesbrough, H., Heaton, S., & Teece, D. J. 2019. Strategic management of open innovation: A dynamic capabilities perspective. *California Management Review*, 62(1): 77-94.

Böhringer, C. & Jochem, P. E. 2007. Measuring the immeasurable—A survey of sustainability indices. *Ecological Economics*, 63(1): 1-8.

Boons, F. & Lüdeke-Freund, F. 2013. Business models for sustainable innovation: State-of-the-art and steps towards a research agenda. *Journal of Cleaner Production*, 45: 9-19.

Bor, S. 2014. *A theory of meta-organisation: An analysis of steering processes in European Commission-funded R&D 'Network of Excellence' consortia.* Helsinki, Finland Hanken School of Economics.

Bosch-Sijtsema, P. M. & Bosch, J. 2015. Plays nice with others? Multiple ecosystems, various roles and divergent engagement models. *Technology Analysis & Strategic Management*, 27(8): 960-974.

Boschma, R. 2015. Towards an evolutionary perspective on regional resilience. *Regional Studies*, 49(5): 733-751.

Boström, M. 2006. Regulatory credibility and authority through inclusiveness: Standardization organizations in cases of eco-labelling. *Organization*, 13(3): 345-367.

Bouwen, R. & Taillieu, T. 2004. Multi-party collaboration as social learning for interdependence: Developing relational knowing for sustainable natural resource management. *Journal of Community & Applied Social Psychology*, 14(3): 137-153.

Boyatzis, R. E. 1998. *Transforming qualitative information: Thematic analysis and code development*. Thousand Oaks, CA: Sage Publications.

Bres, L., Raufflet, E., & Boghossian, J. 2018. Pluralism in organizations: Learning from unconventional forms of organizations. *International Journal of Management Reviews*, 20(2): 364-386.

Bunn, M. D., Savage, G. T., & Holloway, B. B. 2002. Stakeholder analysis for multi-sector innovations. *Journal of Business & Industrial Marketing*, 17(2/3): 181-203.

Burt, R. S. 1992. *Structural holes: The social structure of competition*. Cambridge, MA: Harvard University Press.

Bushe, G. 2012. Feature choice by Gervase Bushe Foundations of appreciative inquiry: History, criticism and potential. *AI Practitioner*, 14(1): 8-20.

Bushe, G. R. & Paranjpey, N. 2015. Comparing the generativity of problem solving and appreciative inquiry: A field experiment. *The Journal of Applied Behavioral Science*, 51(3): 309-335.

Buuren, A., Boons, F., & Teisman, G. 2012. Collaborative problem solving in a complex governance system: Amsterdam airport schiphol and the challenge to break path dependency. *Systems Research & Behavioral Science*, 29(2): 116-130.

Byrne, B. M. 2004. Testing for multigroup invariance using AMOS graphics: A road less traveled. *Structural Equation Modeling*, 11(2): 272-300.

Byrne, B. M. 2016. *Structural equation modeling with AMOS: Basic concepts, applications, and programming* (3rd ed.). New York, NY: Routledge.

Cacioppe, R. & Edwards, M. 2005. Seeking the Holy Grail of organisational development. *Leadership & Organization Development Journal*, 26(2): 86-105.

Cameron, K. S., Dutton, J. E., & Quinn, R. 2003. An introduction to positive organizational scholarship. *Positive Organizational Scholarship*, 3(13): 2-21.

Cameron, K. S. & Spreitzer, G. M. 2011. *The Oxford handbook of positive organizational scholarship*. Oxford: Oxford University Press.

Cañeque, F. C. & Hart, S. L. 2017. *Base of the pyramid 3.0: sustainable development through innovation and entrepreneurship*. New York, NY: Routledge.

Capdevila, I. 2018. Knowing communities and the innovative capacity of cities. *City, Culture and Society*, 13: 8-12.

Carley, M. & Smith, H. 2013. *Urban development and civil society: The role of communities in sustainable cities*. New York, NY: Routledge.

Carrillo-Hermosilla, J., Del Río, P., & Könnölä, T. 2010. Diversity of eco-innovations: Reflections from selected case studies. *Journal of Cleaner Production*, 18(10-11): 1073-1083.

Cassiolato, J. E., Pessoa de Matos, M. G., & Lastres, H. M. 2014. Innovation systems and development. In B. Currie-Alder, R. Kanbur, D. M. Malone, & R. Medhora (Eds.), *International development: Ideas, experience, and prospects*. Oxford, UK: Oxford University Press.

Charmaz, K. 2014. *Constructing grounded theory* (2nd ed.). London: SAGE Publications Ltd.

Chaudhury, A. S., Ventresca, M. J., Thornton, T. F., Helfgott, A., Sova, C., Baral, P., Rasheed, T., & Ligthart, J. 2016. Emerging meta-organisations and adaptation to global climate change: Evidence from implementing adaptation in Nepal, Pakistan and Ghana. *Global Environmental Change-Human and Policy Dimensions*, 38: 243-257.

Cherkowski, S. & Walker, K. 2014. Flourishing communities: Re-storying educational leadership using a positive research lens. *International Journal of Leadership in Education*, 17(2): 200-216.

Choi, J. S. & Pattent, B. C. 2001. Sustainable development: lessons from the paradox of enrichment. *Ecosystem Health*, 7(3): 163-178.

Choi, S., Morgan, S., & Burnett, K. 1998. Phenomenological damping in trapped atomic Bose-Einstein condensates. *Physical Review A*, 57(5): 4057-4060.

Chong, M., Habib, A., Evangelopoulos, N., & Park, H. W. 2018. Dynamic capabilities of a smart city: An innovative approach to discovering urban problems and solutions. *Government Information Quarterly*, 35(4): 682-692.

Ciborra, C. U. 1996. The platform organization: Recombining strategies, structures, and surprises. *Organization Science*, 7(2): 103-118.

Clark, G. 2006. Sustainable development finance for cities and regions. *OECD Papers*, 6(12): 2-14.

Clarke, D., Werestiuk, K., Schoffner, A., Gerard, J., Swan, K., Jackson, B., Steeves, B., & Probizanski, S. 2012. Achieving the perfect handoff' in patient transfers: building teamwork and trust. *Journal of Nursing Management*, 20(5): 592-598.

Clarke, L. 1987. Review of "the managerial imperative: The age of macromanagement" by D. E. Mcfarland. *Contemporary Sociology: A Journal of Reviews*, 16(6): 809-810.

Clarysse, B., Wright, M., Bruneel, J., & Mahajan, A. 2014. Creating value in ecosystems: Crossing the chasm between knowledge and business ecosystems. *Research Policy*, 43(7): 1164-1176.

Clay, L. 2017. *Understanding the factors that lead to successful sustainable development implementation in U.S. cities*: Unpublished Qualitative Research Report, Weatherhead School of Management, Case Western Reserve University, Cleveland, OH.

Cole, M., Stavros, J., & Zerilli, M. 2017. *The SOAR profile (TSP) 2.0: A rapid assessment tool to help individuals and teams build strategic capacity*. Southfield, MI: Fifth Annual Research Day. Lawrence Technological University.

Cole, M. L., Cox, J. D., & Stavros, J. M. 2018. SOAR as a mediator of the relationship between emotional intelligence and collaboration among professionals working in teams: Implications for entrepreneurial teams. *SAGE Open*, 8(2): 2158244018779109.

Cole, M. L., Cox, J. D., & Stavros, J. M. 2019. Building collaboration in teams through emotional intelligence: Mediation by SOAR (strengths, opportunities, aspirations, and results). *Journal of Management & Organization*, 25(2): 263-283.

Cole, M. L. & Stavros, J. M. 2019. SOAR: A framework to build positive psychological capacity in strategic thinking, planning, and leading. In L. E. Van Zyl & S. Rothmann, Sr. (Eds.), *Theoretical approaches to multi-cultural positive psychological interventions*: 505-521. Cham: Springer.

Colombo, M. G. & Delmastro, M. 2002. The determinants of organizational change and structural inertia: Technological and organizational factors. *Journal of Economics & Management Strategy*, 11(4): 595-635.

Conroy, M. M. 2006. Moving the middle ahead: Challenges and opportunities of sustainability in Indiana, Kentucky, and Ohio. *Journal of Planning Education and Research*, 26(1): 18-27.

Cooper, R., Evans, G., & Boyko, C. (Eds.). 2009. *Designing sustainable cities*. Oxford, UK: John Wiley & Sons.

Cooperrider, D., Whitney, D. D., & Stavros, J. 2008. *The appreciative inquiry handbook: For leaders of change* (2nd ed.). San Francisco, CA: Berrett-Koehler Publishers.

Cooperrider, D., Godwin, L., Boland, B., & Avital, M. 2012. The appreciative inquiry summit explorations into the magic of macro-management and crowdsourcing. *AI Practitioner*, 14(2): 4-9.

Cooperrider, D. L., Whitney, D. K., & Stavros, J. M. 2003. *Appreciative inquiry handbook: The first in a series of AI workbooks for leaders of change*. San Francisco, CA: Berrett-Koehler Publishers.

Cooperrider, D. L. & McQuaid, M. 2012. The positive arc of systemic strengths: How appreciative inquiry and sustainable designing can bring out the best in human systems. *Journal of Corporate Citizenship*, 46(Summer 2012): 71-102.

Cooperrider, D. L. & Zhexembayeva, N. 2012. Embedded sustainability and the innovation-producing potential of the UN global compact's environmental principles'. In J. T. Lawrence & P. W. Beamish (Eds.), *Globally responsible leadership: Managing according to the UN Global Compact*: 107-127. Thousand Oaks, CA: SAGE Publications, Inc.

Cooperrrider, D. L. 2012. The concentration effect of strengths. *Organizational Dynamics*, 41(2): 106.

Corbin, J. & Strauss, A. 1990. Grounded theory research: Procedures, canons and evaluative criteria. *Zeitschrift für Soziologie*, 19(6): 418-427.

Corona, B., Shen, L., Reike, D., Carreón, J. R., & Worrell, E. 2019. Towards sustainable development through the circular economy—A review and critical assessment on current circularity metrics. *Resources, Conservation and Recycling*, 151: 104498.

Cozzens, S. & Sutz, J. 2014. Innovation in informal settings: Reflections and proposals for a research agenda. *Innovation and Development*, 4(1): 5-31.

Dale, A., Ling, C., & Newman, L. 2010. Community vitality: The role of community-level resilience adaptation and innovation in sustainable development. *Sustainability*, 2(1): 215-231.

Dale, A., Foon, R., Herbert, Y., & Newell, R. 2015. *Community vitality: From adaptation to transformation*: Fernweh Press.

Dalton, C. M. & Dalton, D. R. 2006. Corporate governance best practices: The proof is in the process. *Journal of Business Strategy*, 27(4): 5-7.

Davidson-Hunt, I. J. 2006. Adaptive learning networks: Developing resource management knowledge through social learning forums. *Human Ecology*, 34(4): 593-614.

Davies, S. R. & Horst, M. 2015. Responsible innovation in the US, UK and Denmark: Governance landscapes. In B.-J. Koops, H. Romijn, I. Oosterlaken, J. van den Hoven, & T. Swierstra (Eds.), *Responsible innovation 2: Concepts, approaches, and applications*: 37-56: Springer.

Davoudi, S., Shaw, K., Haider, L. J., Quinlan, A. E., Peterson, G. D., Wilkinson, C., Fünfgeld, H., McEvoy, D., Porter, L., & Davoudi, S. 2012. Resilience: a bridging concept or a dead end? "Reframing" resilience: Challenges for planning theory and practice interacting traps: resilience assessment of a pasture management system in Northern Afghanistan urban resilience: what does it mean in planning practice? Resilience as a useful concept for climate change adaptation? The politics of resilience for planning: a cautionary note: edited by Simin Davoudi and Libby Porter. *Planning Theory & Practice*, 13(2): 299-333.

de Jong, M., Han, M., & Lu, H. 2019. City branding in Chinese cities: From tactical greenwashing to successful industrial transformation. In X. Zhang (Ed.), *Remaking sustainable urbanism*: 81-99. Singapore: Palgrave Macmillan.

De Saille, S. 2015. Innovating innovation policy: The emergence of 'Responsible Research and Innovation'. *Journal of Responsible Innovation*, 2(2): 152-168.

Delmas, M. A. & Burbano, V. C. 2011. The drivers of greenwashing. *California Management Review*, 54(1): 64-87.

Dollery, B., Crase, L., & Grant, B. J. 2011. The local capacity, local community and local governance dimensions of sustainability in Australian local government. *Commonwealth Journal of Local Governance*, 8-9: 162-183.

Doniger, W. 1999. *Merriam-Webster's encyclopedia of world religions*: Merriam-Webster.

Dunfy, D., Griffiths, A., & Benn, S. 2007. *Organizational change for corporate sustainability: A guide for leaders and change agents for the future*. New York, NY: Routledge.

Dunphy, D. 2000. Embracing paradox: Top-down versus participative management of organizational change, a commentary on Conger and Bennis. In M. Beer & N.

Nohria (Eds.), ***Breaking the code of change***: 99-112: Harvard Business School Press.

Dunphy, D. 2003. Corporate sustainability: Challenge to managerial orthodoxies. ***Journal of Management and Organization***, 9(1): 2-11.

Duxbury, N., Cullen, C., & Pascual, J. 2012. Cities, culture and sustainable development. In H. K. Anheier & Y. R. Isar (Eds.), ***Cultural policy and governance in a new metropolitan age***: 73-86. London: SAGE Publications, Inc.

Edwards, M. G. 2005. The integral holon: A holonomic approach to organisational change and transformation. ***Journal of Organizational Change Management***, 18(3): 269-288.

Edwards, M. G. 2009. An integrative metatheory for organisational learning and sustainability in turbulent times. ***The Learning Organization***, 16(3): 189-207.

Eisenhardt, K. M. & Martin, J. A. 2000. Dynamic capabilities: What are they? ***Strategic Management Journal***, 21(10-11): 1105-1121.

Espey, J., Dahmm, H., & Manderino, L. 2018. ***Leaving no US city behind: The US cities sustainable development goals index***. Paris: Sustainable Development Solutions Network.

Feng, N., Fu, C., Wei, F., Peng, Z., Zhang, Q., & Zhang, K. H. 2019. The key role of dynamic capabilities in the evolutionary process for a startup to develop into an innovation ecosystem leader: An indepth case study. ***Journal of Engineering and Technology Management***, 54: 81-96.

Fernández, N., Maldonado, C., & Gershenson, C. 2014. Information measures of complexity, emergence, self-organization, homeostasis, and autopoiesis. In M. Prokopenko (Ed.), ***Guided self-organization: Inception***: 19-51. Berlin Heidelberg: Springer-Verlag.

Fisher, K., Geenen, J., Jurcevic, M., McClintock, K., & Davis, G. 2009. Applying asset-based community development as a strategy for CSR: A Canadian perspective on a win–win for stakeholders and SMEs. ***Business ethics: A European review***, 18(1): 66-82.

Fiss, P. C. 2011. Building better causal theories: A fuzzy set approach to typologies in organization research. ***Academy of Management Journal***, 54(2): 393-420.

Flora, C. B., Emery, M., Fey, S., & Bregendahl, C. 2005. ***Community capitals: A tool for evaluating strategic interventions and projects***. Ames, IA: North Central Regional Center for Rural Development.

Flora, C. B., Flora, J. L., & Gasteyer, S. P. 2016. *Rural communities: Legacy+ change*: Westview Press.

Florida, R. 2002. The economic geography of talent. *Annals of the Association of American Geographers*, 92(4): 743-755.

Florida, R., Mellander, C., & Stolarick, K. 2008. Inside the black box of regional development—human capital, the creative class and tolerance. *Journal of Economic Geography*, 8(5): 615-649.

Florida, R. 2014. *The rise of the creative class--revisited: Revised and expanded*: Basic Books (AZ).

Florini, A. & Pauli, M. 2018. Collaborative governance for the sustainable development goals. *Asia & the Pacific Policy Studies*, 5(3): 583-598.

Folke, C., Carpenter, S., Elmqvist, T., Gunderson, L., Holling, C. S., & Walker, B. 2002. Resilience and sustainable development: Building adaptive capacity in a world of transformations. *AMBIO: A Journal of the Human Environment*, 31(5): 437-440.

Folke, C., Carpenter, S., Walker, B., Scheffer, M., Chapin, T., & Rockström, J. 2010. Resilience thinking: integrating resilience, adaptability and transformability. *Ecology and Society*, 15(4): 20.

Forrester, J. W. 1970. Urban dynamics. *Industrial Management Review (IMR) (pre-1986)*, 11(3): 67.

Forrester, J. W. 1998. *Designing the future*. Sevilla, Spain: Universidad de Sevilla.

Frantzeskaki, N. 2019. How city-networks are shaping and failing innovations in urban institutions for sustainability and resilience. *Global Policy*, 10(4): 712-714.

Fredrickson, B. L. & Losada, M. F. 2005. Positive affect and the complex dynamics of human flourishing. *American Psychologist*, 60(7): 678-686.

Fredrickson, B. L., Cohn, M. A., Coffey, K. A., Pek, J., & Finkel, S. M. 2008. Open hearts build lives: positive emotions, induced through loving-kindness meditation, build consequential personal resources. *Journal of personality and social psychology*, 95(5): 1045.

Fry, R. 2012. Improving safety in a steel mill words really can create worlds! *AI Practitioner*, 14(2).

Fuller, J., Bartl, M., Ernst, H., & Muhlbacher, H. 2004. Community based innovation: A method to utilize the innovative potential of online communities, *Proceedings of the 37th Annual Hawaii International Conference on System Sciences, 2004*. Big Island, HI: IEEE.

Fuller, R. B. 2008. *Operating manual for spaceship earth*: Estate of R. Buckminster Fuller.

Fung, A. & Wright, E. O. 2003. *Deepening democracy: Institutional innovations in empowered participatory governance*: Verso.

Furaker, B. 2020. European trade union cooperation, union density and employee attitudes to unions. *Transfer-European Review of Labour and Research*, 26(3): 345-358.

Gawer, A. 2014. Bridging differing perspectives on technological platforms: Toward an integrative framework. *Research Policy*, 43(7): 1239-1249.

Gawer, A. & Cusumano, M. A. 2014. Industry platforms and ecosystem innovation. *Journal of Product Innovation Management*, 31(3): 417-433.

George, J. M. & Zhou, J. 2001. When openness to experience and conscientiousness are related to creative behavior: An interactional approach. *Journal of Applied Psychology*, 86(3): 513-524.

Ginsberg, A., Horwitch, M., Mahapatra, S., & Singh, C. 2010. Ecosystem strategies for complex technological innovation: The case of smart grid development, *PICMET 2010 Technology Management for Global Economic Growth*: 1-8. Phuket, Thailand IEEE.

Glasbergen, P. 2002. The green polder model: institutionalizing multi-stakeholder processes in strategic environmental decision-making. *European Environment*, 12(6): 303-315.

Glaser, B. G. & Strauss, A. L. 2009. *The discovery of grounded theory: Strategies for qualitative research*. New Brunswick, NJ: Aldine Transaction.

Glovis, M. J., Cole, M. L., & Stavros, J. M. 2014. SOAR and motivation as mediators of the relationship between flow and project success. *Organization Development Journal*, 32(3): 57-73.

Gowdy, J. M. 1994. The social context of natural capital: the social limits to sustainable development. *International Journal of Social Economics*, 21(8): 43-55.

Greckhamer, T., Furnari, S., Fiss, P. C., & Aguilera, R. V. 2018. Studying configurations with qualitative comparative analysis: Best practices in strategy and organization research. *Strategic Organization*, 16(4): 482-495.

Greene, J. C., Caracelli, V. J., & Graham, W. F. 1989. Toward a conceptual framework for mixed-method evaluation designs. *Educational Evaluation and Policy Analysis*, 11(3): 255-274.

Guizzardi, G. 2012. Ontological foundations for conceptual modeling with applications, *International Conference on Advanced Information Systems Engineering*: 695-696: Springer.

Gulati, R. & Puranam, P. 2009. Renewal through reorganization: The value of inconsistencies between formal and informal organization. *Organization Science*, 20(2): 422-440.

Gulati, R. 2010. *Reorganize for resilience: Putting customers at the center of your business*: Harvard Business Press.

Gulati, R., Puranam, P., & Tushman, M. 2012. Meta-organization design: Rethinking design in interorganizational and community contexts. *Strategic Management Journal*, 33(6): 571-586.

Gulati, R., Mayo, A. J., & Nohria, N. 2016. *Management: An integrated approach*: Cengage Learning.

Hair, J. F., Black, W. C., Babin, B. J., Anderson, R. E., & Tatham, R. L. 1998. *Multivariate data analysis* (5th ed.). Upper Saddle River, NJ: Prentice Hall

Hair, J. F., Risher, J. J., Sarstedt, M., & Ringle, C. M. 2019. When to use and how to report the results of PLS-SEM. *European Business Review*, 31(1): 2-24.

Hair, J. F., Jr., Black, W. C., Babin, B. J., & Anderson, R. E. 2010. *Multivariate data analysis* (7th ed.). Upper Saddle River, NJ: Pearson Prentice Hall.

Hamann, R. & April, K. 2013. On the role and capabilities of collaborative intermediary organisations in urban sustainability transitions. *Journal of Cleaner Production*, 50: 12-21.

Hamel, G. 1998. Opinion: Strategy innovation and the quest for value. *Sloan Management Review*, 39(2): 7-14.

Hansson, S., Arfvidsson, H., & Simon, D. 2019. Governance for sustainable urban development: the double function of SDG indicators. *Area Development and Policy*, 4(3): 217-235.

Hart, S. L. 2017. Prologue: Defining the path towards a BoP 3.0. In F. C. Cañeque & S. L. Hart (Eds.), *Base of the pyramid 3.0: Sustainable development through innovation and entrepreneurship*: 1-4. New York, NY: Routledge.

Hartmann, A. 1991. Words create worlds. *Social Work*, 36(4): 275-276.

Hawkins, D. R. 2014. *Power vs. force: The hidden determinants of human behavior*: Hay House Incorporated.

Heaton, S., Siegel, D. S., & Teece, D. J. 2019. Universities and innovation ecosystems: A dynamic capabilities perspective. *Industrial and Corporate Change*, 28(4): 921-939.

Helfat, C. E., Finkelstein, S., Mitchell, W., Peteraf, M., Singh, H., Teece, D., & Winter, S. G. 2009. *Dynamic capabilities: Understanding strategic change in organizations*: John Wiley & Sons.

Helfat, C. E. & Raubitschek, R. S. 2018. Dynamic and integrative capabilities for profiting from innovation in digital platform-based ecosystems. *Research Policy*, 47(8): 1391-1399.

Hemmati, M. 2002. *Multi-stakeholder processes for governance and sustainability: Beyond deadlock and conflict*: Routledge.

Henrekson, M. 2014. Entrepreneurship, innovation, and human flourishing. *Small Business Economics*, 43(3): 511-528.

Herrmann, T., Jahnke, I., & Loser, K.-U. 2004. The role concept as a basis for designing community systems. In F. Darses, R. Dieng, C. Simone, & M. Zackland (Eds.), *Cooperative systems design: Scenario-based design of collaborative systems*: 163-178. Amsterdam: IOS Press.

Holling, C. S. 2001. Understanding the complexity of economic, ecological, and social systems. *Ecosystems*, 4(5): 390-405.

Holm, J. R., Lorenz, E., Lundvall, B.-Å., & Valeyre, A. 2010. Organizational learning and systems of labor market regulation in Europe. *Industrial and Corporate Change*, 19(4): 1141-1173.

Homsy, G. C. & Warner, M. E. 2015. Cities and sustainability: Polycentric action and multilevel governanc. *Urban Affairs Review*, 51(1): 46-73.

Hu, L. t. & Bentler, P. M. 1999. Cutoff criteria for fit indexes in covariance structure analysis: Conventional criteria versus new alternatives. *Structural Equation Modeling: A Multidisciplinary Journal*, 6(1): 1-55.

Hurmelinna-Laukkanen, P. & Nätti, S. 2018. Orchestrator types, roles and capabilities–A framework for innovation networks. *Industrial Marketing Management*, 74: 65-78.

Ingraham, P. W., Joyce, P. G., & Donahue, A. K. 2003. *Government performance: Why management matters*: Taylor & Francis.

Jackson, T. 2016. *Prosperity without growth: Foundations for the economy of tomorrow*: Taylor & Francis.

Jensen, P. E. 2005. A contextual theory of learning and the learning organization. *Knowledge and Process Management*, 12(1): 53-64.

Jepson, E. J., Jr. 2004. The adoption of sustainable development policies and techniques in US cities: How wide, how deep, and what role for planners? *Journal of Planning Education and Research*, 23(3): 229-241.

Johnson, R. B. & Onwuegbuzie, A. J. 2004. Mixed methods research: A research paradigm whose time has come. *Educational Researcher*, 33(7): 14-26.

Jonas, J. M., Boha, J., Sörhammar, D., & Moeslein, K. M. 2018. Stakeholder engagement in intra-and inter-organizational innovation: Exploring antecedents of engagement in service ecosystems. *Journal of Service Management*, 29(3): 399-421.

Kapoor, R. & Lee, J. M. 2013. Coordinating and competing in ecosystems: How organizational forms shape new technology investments. *Strategic Management Journal*, 34(3): 274-296.

Kasemir, B., van Asselt, M. B., Durrenberger, G., & Jaeger, C. C. 1999. Integrated assessment of sustainable development: Multiple perspectives in interaction. *International Journal of Environment and Pollution*, 11(4): 407-425.

Katz, B. & Wagner, J.; The rise of urban innovation districts; https://hbr.org/2014/11/the-rise-of-urban-innovation-districts.

Kay, J. J. & Schneider, E. 1995. Embracing complexity the challenge of the ecosystem approach, *Perspectives On Ecological Integrity*: 49-59: Springer.

Kay, J. J., Regier, H. A., Boyle, M., & Francis, G. 1999. An ecosystem approach for sustainability: Addressing the challenge of complexity. *Futures*, 31(7): 721-742.

Kemp, R. & Loorbach, D. 2003. Governance for sustainability through transition management, *Open Meeting of Human Dimensions of Global Environmental Change Research Community*: 16-18. Montreal, Canada.

Kemp, R. & Rotmans, J. 2005. The management of the co-evolution of technical, environmental and social systems. In J. Hemmelskamp & K. M. Weber (Eds.), *Towards environmental innovation systems*: 33-55: Springer.

Khavarian-Garmsir, A. R. & Zare, S. M. 2015. SOAR framework as a new model for the strategic planning of sustainable tourism. *Tourism Planning & Development*, 12(3): 321-332.

Kirkman, B. L., Mathieu, J. E., Cordery, J. L., Rosen, B., & Kukenberger, M. 2011. Managing a new collaborative entity in business organizations: Understanding organizational communities of practice effectiveness. *Journal of Applied Psychology*, 96(6): 1234-1245.

Klein, K. J., Tosi, H., & Cannella, A. A., Jr. 1999. Multilevel theory building: Benefits, barriers, and new developments. *Academy of Management Review*, 24(2): 248-253.

Kline, R. 2011. *Principles and practice of structural equation modeling* (3rd ed.). New York, NY: Guilford Press.

Koops, B.-J. 2015. The concepts, approaches, and applications of responsible innovation. In B.-J. Koops, I. Oosterlaken, H. Romijn, T. Swierstra, & J. van den Hoven (Eds.), *Responsible Innovation 2*: 1-15: Springer.

Laszlo, C. & Cooperrider, D. L. 2010. Creating sustainable value: A strength-based whole system approach. In T. Thatchenkery, D. Cooperrider, & M. Avital (Eds.), *Positive design and appreciative construction: From sustainable development to sustainable value*, vol. 4: 17-33: Emerald Group Publishing Limited.

Laszlo, C. & Brown, J. S. 2014. *Flourishing enterprise: The new spirit of business*: Stanford University Press.

Laszlo, C. & Zhexembayeva, N. 2017. *Embedded sustainability: The next big competitive advantage*: Routledge.

Laszlo, C. 2020. Quantum management: the practices and science of flourishing enterprise. *Journal of Management, Spirituality & Religion*, 17(4): 301-315.

Laurent, A., Garaudel, P., Schmidt, G., & Eynaud, P. 2020. Civil society meta-organizations and legitimating processes: The case of the addiction field in France. *Voluntas*, 31(1): 19-38.

Lawson, C. & Lorenz, E. 1999. Collective learning, tacit knowledge and regional innovative capacity. *Regional Studies*, 33(4): 305-317.

Lawton, T. C., Rajwani, T., & Minto, A. 2018. Why trade associations matter: Exploring function, meaning, and influence. *Journal of Management Inquiry*, 27(1): 5-9.

Lee, R. G. & Petts, J. 2013. Adaptive governance for responsible innovation. In J. R. Bessant, M. Heintz, & R. Owen (Eds.), *Responsible innovation: Managing the responsible emergence of science and innovation in society*: 143-164: Wiley.

Lester, S. M. 2002. *The tools of government. A guide to the New Governance*. New York, NY: Oxford University Press.

Li, J. F. & Garnsey, E. 2014. Building joint value: Ecosystem support for global health innovations. In R. Adner, J. E. Oxley, & B. S. Silverman (Eds.), *Collaboration and competition in business ecosystems*: 69-98: Emerald Group Publishing Limited.

Libbrecht, N., Lievens, F., & Schollaert, E. 2010. Measurement equivalence of the Wong and Law Emotional Intelligence Scale across self and other ratings. *Educational and Psychological Measurement*, 70(6): 1007-1020.

Lindblom, C. E. 1999. A century of planning, *Planning sustainability*: 39-65: Routledge.

Linde, L., Sjödin, D., Parida, V., & Wincent, J. 2021. Dynamic capabilities for ecosystem orchestration A capability-based framework for smart city innovation initiatives. *Technological Forecasting and Social Change*, 166: 120614.

Linnér, B.-O. & Wibeck, V. 2020. Conceptualising variations in societal transformations towards sustainability. *Environmental Science & Policy*, 106: 221-227.

Lubell, M., Feiock, R., & Handy, S. 2009a. City adoption of environmentally sustainable policies in California's Central Valley. *Journal of the American Planning Association*, 75(3): 293-308.

Lubell, M., Leach, W. D., & Sabatier, P. A. 2009b. Collaborative watershed partnerships in the epoch of sustainability. In D. A. Mazmanian & M. E. Kraft (Eds.), *Toward sustainable communities: Transitions and transformations in environmental policy*, 2nd ed.: 255-288. Cambridge, MA: The MIT Press.

Luck, G. W. 2007. A review of the relationships between human population density and biodiversity. *Biological Reviews*, 82(4): 607-645.

Luther, V. & Flora, C. B. n.d. Building community capacity: https://www.researchgate.net/publication/288902923_An_introduction_to_building_community_capacity.

Lütjen, H., Schultz, C., Tietze, F., & Urmetzer, F. 2019. Managing ecosystems for service innovation: A dynamic capability view. *Journal of Business Research*, 104: 506-519.

Lynch, A. 2019. *2019 US cities sustainable development report*: Sustainable Development Solutions Network. https://www.sustainabledevelopment.report/reports/2019-us-cities-sustainable-development-report/.

MacCoy, D. J. 2014. Appreciative inquiry and evaluation–Getting to what works. *Canadian Journal of Program Evaluation*, 29(2): 104-127.

MacDonald, A., Clarke, A., Ordonez-Ponce, E., Chai, Z., & Andreasen, J. 2020. Sustainability managers: The job roles and competencies of building sustainable cities and communities. *Public Performance & Management Review*, 43(6): 1413-1444.

Macellari, M., Yuriev, A., Testa, F., & Boiral, O. 2021. Exploring bluewashing practices of alleged sustainability leaders through a counter-accounting analysis. *Environmental Impact Assessment Review*, 86: 106489.

Macharis, C. & Kin, B. 2017. The 4 A's of sustainable city distribution: Innovative solutions and challenges ahead. *International Journal of Sustainable Transportation*, 11(2): 59-71.

Maier, S. F. & Seligman, M. E. 2016. Learned helplessness at fifty: Insights from neuroscience. *Psychological Review*, 123(4): 349-367.

Malcourant, E., Vas, A., & Zintz, T. 2015. World Anti-Doping Agency: A meta-organizational perspective. *Sport, Business and Management: An International Journal*, 5(5): 451-471.

Malhotra, N. K., Mukhopadhyay, S., Liu, X., & Dash, S. 2012. One, few or many?: an integrated framework for identifying the items in measurement scales. *International Journal of Market Research*, 54(6): 835-862.

Marsal-Llacuna, M.-L., Colomer-Llinàs, J., & Meléndez-Frigola, J. 2015. Lessons in urban monitoring taken from sustainable and livable cities to better address the Smart Cities initiative. *Technological Forecasting and Social Change*, 90: 611-622.

Matsunaga, M. 2010. How to Factor-Analyze Your Data Right: Do's, Don'ts, and How-To's. *International Journal of Psychological Research*, 3(1).

Maturana, H. R., Varela, F. J., Uribe, R. B., & Zeleny, M. 1981. A special edition devoted to autopoiesis. *Cybernetics Forum*, X(2 and 3): 1-38.

Maxton, P. J. & Bushe, G. R. 2018. Individual cognitive effort and cognitive transition during organization development. *The Journal of Applied Behavioral Science*, 54(4): 424-456.

McClellan, J. L. 2007. Marrying positive psychology to mediation: Using appreciative inquiry and solution-focused counseling to improve the process. *Dispute Resolution Journal*, 62(4): 29.

McDonald, R. P. & Marsh, H. W. 1990. Choosing a multivariate model: Noncentrality and goodness of fit. *Psychological Bulletin*, 107(2): 247.

McFarland, D. E. 1974. Management and its critics: A look at social pluralism. *Journal of Business Research*, 2(4): 395-407.

McFarland, D. E. 1977. Management, humanism, and society: The case for macromanagement theory. *Academy of Management Review*, 2(4): 613-623.

McKnight, J. 2017. *Asset-based community development: The essentials*. Chicago: Asset-Based Community Development Institute.

McQuaid, M. 2019. *How do appreciative inquiry summits help organizational systems to flourish?*: University of Twente.

Meadows, D. 1999. Indicators and information systems for sustainable development. In D. Satterthwaite (Ed.), *The Earthscan reader in sustainable cities*: 364-393. New York, NY: Routledge.

Misangyi, V. F., Greckhamer, T., Furnari, S., Fiss, P. C., Crilly, D., & Aguilera, R. 2017. Embracing causal complexity: The emergence of a neo-configurational perspective. *Journal of Management*, 43(1): 255-282.

Mitchell, M. 2009. *Complexity: A guided tour*: Oxford University Press.

Mol, A. P. 2000. The environmental movement in an era of ecological modernisation. *Geoforum*, 31(1): 45-56.

Morley, C.; All cities are mad: but the madness is gallant. All cities are beautiful: but the beauty is grim [Christopher Morley, Where the Blue Begins (1922)] https://www.brainyquote.com/quotes/christopher_morley_386584.

Mousavi, S., Bossink, B., & van Vliet, M. 2018. Dynamic capabilities and organizational routines for managing innovation towards sustainability. *Journal of Cleaner Production*, 203: 224-239.

Muthén, B. 2011. Applications of causally defined direct and indirect effects in mediation analysis using SEM in Mplus: http://www.statmodel.com/download/causalmediation.pdf.

Na, S., Chung, D., Jung, J., Hula, A., Fiore, V. G., Dayan, P., & Gu, X. 2019. *Humans use forward thinking to exert social control*: Available at SSRN: https://ssrn.com/abstract=3443690 or http://dx.doi.org/10.2139/ssrn.3443690

Neary, J. & Osborne, M. 2018. University Engagement in Achieving Sustainable Development Goals: A Synthesis of Case Studies from the SUEUAA Study. *Australian Journal of Adult Learning*, 58(3): 336-364.

Nel, H. 2015. An integration of the livelihoods and asset-based community development approaches: A South African case study. *Development Southern Africa*, 32(4): 511-525.

Nel, H. 2018. Community leadership: A comparison between asset-based community-led development (ABCD) and the traditional needs-based approach. *Development Southern Africa*, 35(6): 839-851.

Newman, P. & Jennings, I. 2012. *Cities as sustainable ecosystems: Principles and practices*: Island Press.

Nikkhah, H. A. & Redzuan, M. 2009. Participation as a medium of empowerment in community development. *European Journal of Social Sciences*, 11(1): 170-176.

O'Leary, R., Fiorino, D. J., Durant, R. F., & Weiland, P. S. 1999. *Managing for the environment understanding the legal, organizational, and policy challenges*. San Francisco, CA: Jossey-Bass.

Oh, D.-S., Phillips, F., Park, S., & Lee, E. 2016. Innovation ecosystems: A critical examination. *Technovation*, 54: 1-6.

Olawumi, T. O., Chan, D. W., Wong, J. K., & Chan, A. P. 2018. Barriers to the integration of BIM and sustainability practices in construction projects: A Delphi survey of international experts. *Journal of Building Engineering*, 20: 60-71.

Oliveira, E. A., Andrade, J. S., & Makse, H. A. 2014. Large cities are less green. *Scientific Reports*, 4(1): 1-13.

Opdyke, A., Javernick-Will, A., & Koschmann, M. 2018. A comparative analysis of coordination, participation, and training in post-disaster shelter projects. *Sustainability*, 10(11): 4241.

Ostrom, E. 2010. Beyond markets and states: Polycentric governance of complex economic systems. *American Economic Review*, 100(3): 641-672.

Pahl-Wostl, C. 2009. A conceptual framework for analysing adaptive capacity and multi-level learning processes in resource governance regimes. *Global Environmental Change*, 19(3): 354-365.

Pandey, A. & Gupta, R. K. 2008. A perspective of collective consciousness of business organizations. *Journal of Business Ethics*, 80(4): 889-898.

Parris, T. M. & Kates, R. W. 2003. Characterizing and measuring sustainable development. *Annual Review of Environment and Resources*, 28(1): 559-586.

Parrish, B. D. 2010. Sustainability-driven entrepreneurship: Principles of organization design. *Journal of Business Venturing*, 25(5): 510-523.

Pavez, I., Kendall, L. D., & Laszlo, C. 2020. Positive-impact companies: Toward a new paradigm of value creation. *Organizational Dynamics*: 1-10.

Pavlou, P. A. & El Sawy, O. A. 2011. Understanding the elusive black box of dynamic capabilities. *Decision Sciences*, 42(1): 239-273.

Peter, C. & Swilling, M. 2014. Linking complexity and sustainability theories: Implications for modeling sustainability transitions. *Sustainability*, 6(3): 1594-1622.

Peterson, R. A. & Kim, Y. 2013. On the relationship between coefficient alpha and composite reliability. *Journal of Applied Psychology*, 98(1): 194-198.

Podsakoff, P. M., MacKenzie, S. B., & Podsakoff, N. P. 2012. Sources of method bias in social science research and recommendations on how to control it. *Annual Review of Psychology*, 63: 539-569.

Popkov, Y. S. 1984. Macromanagement of spatial models of urban systems. *Environment and Planning A*, 16(12): 1544-1546.

Portney, K. 2005. Civic engagement and sustainable cities in the United States. *Public Administration Review*, 65(5): 579-591.

Portney, K. E. 2003. *Taking sustainable cities seriously: Economic development, the environment, and quality of life in American cities*. Cambridge, MA: MIT Press.

Portney, K. E. & Cuttler, Z. 2010. The local nonprofit sector and the pursuit of sustainability in American cities: A preliminary exploration. *Local Environment*, 15(4): 323-339.

Portney, K. E. 2011. Toward the healthy city: People, places, and the politics of urban planning - By Jason Corburn. Breakthrough Communities: Sustainability and Justice in the Next American Metropolis - Edited by M. Paloma Pavel. *Review of Policy Research*, 28(3): 310-315.

Prakash, M., Teksoz, K., Espey, J., Sachs, J., Shank, M., & Schmidt-Traub, G. 2017. *Achieving a sustainable urban America*: The U.S. Cities Sustainable Development Goals Index 2017. Sustainable Development Solutions Network. https://s3.amazonaws.com/sustainabledevelopment.report/2017/2017_sdg_index_us_cities.pdf.

Preacher, K. J. & Hayes, A. F. 2008. Asymptotic and resampling strategies for assessing and comparing indirect effects in multiple mediator models. *Behavior Research Methods*, 40(3): 879-891.

Priem, R. L. & Butler, J. E. 2001. Is the resource-based "view" a useful perspective for strategic management research? *Academy of Management Review*, 26(1): 22-40.

Prigogine, I. & Stengers, I. 1997. *The end of certainty*: Simon and Schuster.

Prugh, T., Costanza, R., & Daly, H. E. 2000. *The local politics of global sustainability*: Island Press.

Quinney, L., Dwyer, T., & Chapman, Y. 2016. Who, where, and how of interviewing peers: Implications for a phenomenological study. *Sage Open*, 6(3): 2158244016659688.

Ragin, C. C. 1987. *The comparative method: Moving beyond qualitative and quantitative strategies*. Berkeley and Los Angeles: University of California Press.

Ragin, C. C. 2006. Set relations in social research: Evaluating their consistency and coverage. *Political Analysis*, 14(3): 291-310.

Ragin, C. C. 2014. *The comparative method: Moving beyond qualitative and quantitative strategies* (2nd ed.). Oakland, CA: University of California Press.

Rainey, H. G. 2009. *Understanding and managing public organizations*: John Wiley & Sons.

Rajabi, A. & Hesarinejad, J. 2013. Urban non-governmental companies: A solution for citizens' participation in urban management. *Journal of Urban Economics and Management*, 1(4): 111-129.

Rajwani, T., Lawton, T., & Phillips, N. 2015. The "Voice of Industry": Why management researchers should pay more attention to trade associations. *Strategic Organization*, 13(3): 224-232.

Rant, M. B. 2020. Sustainable development goals (SDGs), leadership, and Sadhguru: Self-transformation becoming the aim of leadership development. *The International Journal of Management Education*, 18(3): 100426.

Rasche, A., Waddock, S., & McIntosh, M. 2013. The United Nations global compact: Retrospect and prospect. *Business & Society*, 52(1): 6-30.

Ravetz, J. 1999. Citizen participation for integrated assessment: New pathways in complex systems. *International Journal of Environment and Pollution*, 11(3): 331-350.

Reed, M. S., Fraser, E. D., & Dougill, A. J. 2006. An adaptive learning process for developing and applying sustainability indicators with local communities. *Ecological Economics*, 59(4): 406-418.

Rihoux, B. & Ragin, C. C. 2008. *Configurational comparative methods: Qualitative comparative analysis (QCA) and related techniques*: Sage Publications.

Rihoux, B. & De Meur, G. 2009. Crisp-set qualitative comparative analysis (csQCA). *Configurational comparative methods: Qualitative comparative analysis (QCA) and related techniques*, 51: 33-68.

Rindova, V. P. & Kotha, S. 2001. Continuous "morphing": Competing through dynamic capabilities, form, and function. *Academy of Management Journal*, 44(6): 1263-1280.

Ritzer, G. 1988. Sociological metatheory: a defense of a subfield by a delineation of its parameters. *Sociological Theory*, 6(2): 187-200.

Roepke, A. M. & Seligman, M. E. 2015. Doors opening: A mechanism for growth after adversity. *The Journal of Positive Psychology*, 10(2): 107-115.

Roling, N. G. & Wagemakers, M. A. E. 2000. *Facilitating sustainable agriculture: participatory learning and adaptive management in times of environmental uncertainty*: Cambridge University Press.

Romanelli, M. 2018. Towards Sustainable Ecosystems. *Systems Research & Behavioral Science*, 35(4): 417-426.

Rosado, C. 2008. Context determines content: Quantum physics as a framework for wholeness in urban transformation. *Urban Studies*, 45(10): 2075-2097.

Rousseau, H. E., Berrone, P., & Gelabert, L. 2019. Localizing sustainable development goals: Nonprofit density and city sustainability. *Academy of Management Discoveries*, 5(4): 487-513.

Rousseau, J.-J.; Cities are the abyss of the human species [Jean-Jacques Rousseau, Émile (1762)] https://www.laphamsquarterly.org/contributors/rousseau.

Runhaar, H., Driessen, P. P., & Soer, L. 2009. Sustainable urban development and the challenge of policy integration: an assessment of planning tools for integrating spatial and environmental planning in the Netherlands. *Environment and Planning B: Planning and Design*, 36(3): 417-431.

Ruse, M. 2005. Entry for "reductionism", *The Oxford companion to philosophy*, 2nd ed. Oxford: Oxford University Press.

Sachs, J. D., Schmidt-Traub, G., Mazzucato, M., Messner, D., Nakicenovic, N., & Rockström, J. 2019. Six transformations to achieve the sustainable development goals. *Nature Sustainability*, 2(9): 805-814.

Sachs, J. D., Schmidt-Traub, G., Kroll, C., Lafortune, G., Fuller, G., & Woelm, F. 2020. *The sustainable development goals and COVID-19. Sustainable Development Report 2020*. Cambridge: Cambridge University Press.

Saha, D. 2008. *Sustainable cities: Agenda setting and implementation of sustainability initiatives in US cities*. Unpublished book, The University of Texas at Austin, Austin, TX.

Saha, D. & Paterson, R. G. 2008. Local government efforts to promote the "three Es" of sustainable development. *Journal of Planning Education and Research*, 28(1): 21-37.

Saha, D. 2009. Factors influencing local government sustainability efforts. *State and Local Government Review*, 41(1): 39-48.

Salarpour, H., Amiri, G. G., & Mousavi, S. M. 2019. Criteria assessment in sustainable macromanagement of housing provision problem by a multi-phase decision approach with DEMATEL and dynamic uncertainty. *Arabian Journal for Science and Engineering*, 44(8): 7313-7333.

Saldaña, J. 2021. *The coding manual for qualitative researchers*: Sage.

Sasaki, R. 2014. *Urban prosperity without growth? Sustainable city development with focus on human flourishing*. Unpublished Master's, LUCSUS (Lund University Centre for Sustainability Studies).

Satterthwaite, D. 2010. The role of cities in sustainable development. *Sustainable Development Insights*, 4: 1-8.

Scartezzini, R., Germani, L., & Gritti, R. 1985. *I limiti della democrazia: Autoritarismo e democrazia nella società moderna*: Liguori.

Schaefer, K., Corner, P. D., & Kearins, K. 2015. Social, environmental and sustainable entrepreneurship research: What is needed for sustainability-as-flourishing? *Organization & Environment*, 28(4): 394-413.

Schoemaker, P. J. H., Heaton, S., & Teece, D. 2018. Innovation, dynamic capabilities, and leadership. *California Management Review*, 61(1): 15-42.

Schoenberger-Orgad, M. & McKie, D. 2005. Sustaining edges: CSR, postmodern play, and SMEs. *Public Relations Review*, 31(4): 578-583.

Schuetze, T. & Chelleri, L. 2016. Urban sustainability versus green-washing—Fallacy and reality of urban regeneration in downtown Seoul. *Sustainability*, 8(1): 33.

Scott, S. G. & Bruce, R. 1994a. Creating innovative behavior among R&D professionals: the moderating effect of leadership on the relationship between problem-solving style and innovation, *Proceedings of 1994 IEEE International Engineering Management Conference-IEMC'94*: 48-55: IEEE.

Scott, S. G. & Bruce, R. A. 1994b. Determinants of innovative behavior: A path model of individual innovation in the workplace. *Academy of Management Journal*, 37(3): 580-607.

Seligman, M. E., Railton, P., Baumeister, R. F., & Sripada, C. 2013. Navigating into the future or driven by the past. *Perspectives on Psychological Science*, 8(2): 119-141.

Senge, P. 1990. *The fifth discipline: The art & practice of the learning organization*. New York, NY: Doubleday/Currency.

Senge, P. M. 2003. Taking personal change seriously: The impact of organizational learning on management practice. *Academy of Management Perspectives*, 17(2): 47-50.

Silvestre, B. S. & Țîrcă, D. M. 2019. Innovations for sustainable development: Moving toward a sustainable future. *Journal of Cleaner Production*, 208: 325-332.

Simmie, J. & Martin, R. 2010. The economic resilience of regions: Towards an evolutionary approach. *Cambridge Journal of Regions, Economy and Society*, 3(1): 27-43.

Simon, H. A. 2019. *The sciences of the artificial* (3rd ed.). Cambridge, MA: MIT Press.

Sperry, R. W. 1993. A mentalist view of consciousness. *Social Neuroscience Bulletin*, 6(2): 15-19.

Spillman, L. 2018. Meta-Organization Matters. *Journal of Management Inquiry*, 27(1): 16-20.

Sroufe, R. 2017. Integration and organizational change towards sustainability. *Journal of Cleaner Production*, 162: 315-329.

Sroufe, R. 2018. *Integrated management: How sustainability creates value for any business*: Emerald Group Publishing.

Stacey, R. 2002. Organizations as complex responsive processes of relating. *Journal of Innovative Management*, 8(2): 37-39.

Stacey, R. D. & Griffin, D. 2005. *A complexity perspective on researching organizations: Taking experience seriously*: Taylor & Francis.

Stahl, B. C., Eden, G., & Jirotka, M. 2013. Responsible research and innovation in information and communication technology: Identifying and engaging with the ethical implications of ICTs. *Responsible innovation*: 199-218.

Stark, D. 2009. Ambiguous assets for uncertain environments: Heterarchy in postsocialist firms, *The twenty-first-century firm*: 69-104: Princeton University Press.

Stavros, J., Cooperrider, D., & Kelley, L. 2007. SOAR: A new approach to strategic planning. In P. Holman, T. Devane, & S. Cady (Eds.), *The change handbook*, vol. 3 of 5: 268-284: CreateSpace.

Stavros, J. M. & Hinrichs, G. 2009. *The thin book of SOAR: Building strengths-based strategy*: Thin Book Publishing.

Stavros, J. M. & Saint, D. 2010. SOAR: Linking strategy and OD to sustainable performance. In A. Sullivan, J. M. Stavros, R. L. Sullivan, & W. J. Rothwell (Eds.), *Practicing organization development: A guide for leading change*, 3rd ed.: 377-394. San Francisco, CA: Pfeiffer.

Stavros, J. M., Torres, C., & Cooperrider, D. L. 2018. *Conversations worth having: Using appreciative inquiry to fuel productive and meaningful engagement*: Berrett-Koehler Publishers.

Steele, W. 2017. Planning wild cities, *Connections*: 179-188: Routledge.

Strauss, A. & Corbin, J. 1990. *Basics of qualitative research: Grounded theory procedures and techniques*: SAGE Publications, Inc.

Strauss, A. & Corbin, J. 1994. Grounded theory methodology. In N. K. Denzin & Y. S. Lincoln (Eds.), *Handbook of qualitative research*, vol. 17: 273-285: Sage Publications, Inc.

Streeck, W. & Schmitter, P. C. 1985. Community, market, state-and associations? The prospective contribution of interest governance to social order. *European Sociological Review*, 1(2): 119-138.

Sur, M. 2017. The blue urban: colouring and constructing Kolkata. *City*, 21(5): 597-606.

Tashakkori, A. & Creswell, J. W. 2007. *The new era of mixed methods*: Sage Publications.

Teddlie, C. & Tashakkori, A. 2006. A general typology of research designs featuring mixed methods. *Research in the Schools*, 13(1): 12-28.

Teece, D. J., Pisano, G., & Shuen, A. 1997. Dynamic capabilities and strategic management. *Strategic Management Journal*, 18(7): 509-533.

Teece, D. J. 2007. Explicating dynamic capabilities: the nature and microfoundations of (sustainable) enterprise performance. *Strategic Management Journal*, 28(13): 1319-1350.

Teece, D. J. 2011. Achieving integration of the business school curriculum using the dynamic capabilities framework. *Journal of Management Development*, 30(5): 499-518.

Teece, D. J. & Leih, S. 2016. Uncertainty, innovation, and dynamic capabilities: An introduction. *California Management Review*, 58(4): 5-12.

Teece, D. J. 2018a. Managing the university: Why "organized anarchy" is unacceptable in the age of massive open online courses. *Strategic Organization*, 16(1): 92-102.

Teece, D. J. 2018b. Dynamic capabilities as (workable) management systems theory. *Journal of Management & Organization*, 24(3): 359-368.

Teece, D. J. 2020. Hand in glove: open innovation and the dynamic capabilities framework. *Strategic Management Review*, 1(2): 233-253.

Thabrew, L., Wiek, A., & Ries, R. 2009. Environmental decision making in multi-stakeholder contexts: Applicability of life cycle thinking in development planning and implementation. *Journal of Cleaner Production*, 17(1): 67-76.

Tsao, F. C. & Laszlo, C. 2019. *Quantum leadership: New consciousness in business*: Stanford University Press.

Tsou, H.-T. & Chen, J.-S. 2020. Dynamic capabilities, human capital and service innovation: the case of Taiwan ICT industry. *Asian Journal of Technology Innovation*, 28(2): 181-203.

Tufte, E. R. 1985. The visual display of quantitative information. *The Journal for Healthcare Quality (JHQ)*, 7(3): 15.

Tyre, M. J. & Von Hippel, E. 1997. The situated nature of adaptive learning in organizations. *Organization Science*, 8(1): 71-83.

Valente, M. & Oliver, C. 2018. Meta-organization formation and sustainability in Sub-Saharan Africa. *Organization Science*, 29(4): 678-701.

Van der Zee, K., Thijs, M., & Schakel, L. 2002. The relationship of emotional intelligence with academic intelligence and the Big Five. *European Journal of Personality*, 16(2): 103-125.

Van Gigch, J. P. & Le Moigne, J. L. 1989. A paradigmatic approach to the discipline of information systems. *Behavioral Science*, 34(2): 128-147.

Van Marrewijk, M. & Hardjono, T. W. 2003. European corporate sustainability framework for managing complexity and corporate transformation. *Journal of Business Ethics*, 44(2): 121-132.

Van Marrewijk, M. & Werre, M. 2003. Multiple levels of corporate sustainability. *Journal of Business ethics*, 44(2): 107-119.

Van Marrewijk, M. & Becker, H. M. 2004. The hidden hand of cultural governance: The transformation process of humanitas, a community-driven organization providing, cure, care, housing and well-being to elderly people. *Journal of Business Ethics*, 55(2): 205-214.

Van Waarden, F. 1992. Emergence and development of business interest associations. An example from The Netherlands. *Organization Studies*, 13(4): 521-561.

Varro, M. T.; Divine Nature gave the fields, human art built the cities [Marcus Terentius Varro, De Re Rustica (116 BC–227 BC)]; https://www.laphamsquarterly.org/city/quote/marcus-terentius-varro.

Verhoeven, K. & Maritz, A. 2012. Collaboration for Innovation: Network processes and capabilities, *ISPIM Conference Proceedings*: 1: The International Society for Professional Innovation Management (ISPIM).

Visnjic, I., Neely, A., Cennamo, C., & Visnjic, N. 2016. Governing the city: Unleashing value from the business ecosystem. *California Management Review*, 59(1): 109-140.

Voegtlin, C. & Scherer, A. G. 2017. Responsible innovation and the innovation of responsibility: Governing sustainable development in a globalized world. *Journal of Business Ethics*, 143(2): 227-243.

von Hippel, E. 2009. Democratizing innovation: The evolving phenomenon of user innovation. *International Journal of Innovation Science*, 1(1): 29-40.

Von Schomberg, R. 2011. *Towards responsible research and innovation in the information and communication technologies and security technologies fields*: Available at SSRN: https://ssrn.com/abstract=2436399

Wagner, D. G. & Berger, J. 1985. Do sociological theories grow? *American Journal of Sociology*, 90(4): 697-728.

Walker, T. 2013. The integrative management approach of CSR. In S. O. Idowu, N. Capaldi, L. Zu, & A. Das Gupta (Eds.), *Encyclopedia of corporate social responsibility*: 1712. Berlin: Springer.

Wang, X., Hawkins, C. V., Lebredo, N., & Berman, E. M. 2012. Capacity to sustain sustainability: A study of US cities. *Public Administration Review*, 72(6): 841-853.

Watene, K. & Yap, M. 2015. Culture and sustainable development: Indigenous contributions. *Journal of Global Ethics*, 11(1): 51-55.

Weber, E. P., Lovrich, N. P., & Gaffney, M. 2005. Collaboration, enforcement, and endangered species: A framework for assessing collaborative problem-solving capacity. *Society and Natural Resources*, 18(8): 677-698.

Weber, E. P. & Khademian, A. M. 2008. Wicked problems, knowledge challenges, and collaborative capacity builders in network settings. *Public Administration Review*, 68(2): 334-349.

Weidner, H. 2002. Capacity building for ecological modernization: lessons from cross-national research. *American Behavioral Scientist*, 45(9): 1340-1368.

Weidner, H. & Jänicke, M. (Eds.). 2002. *Summary: Environmental capacity building in a converging world*. Berlin: Springer.

Wells, A. 1970. *Social institutions*: Heinemann.

Werhane, P. H., Hartman, L. P., Archer, C., Englehardt, E. E., & Pritchard, M. S. 2013. *Obstacles to ethical decision-making: Mental models, Milgram and the problem of obedience*: Cambridge University Press.

Westley, F. & McGowan, K. 2017. *The evolution of social innovation: building resilience through transitions*: Edward Elgar Publishing.

Weymouth, R. & Hartz-Karp, J. 2018. Principles for integrating the implementation of the sustainable development goals in cities. *Urban Science*, 2(3): 77.

Whitney, D. 2012. Macro management of meaning and identity explorations into the magic of macro-management and crowdsourcing. *AI Practitioner*, 14(2): 24-28.

Whitney, D. & Fredrickson, B. L. 2015. Appreciative Inquiry Meets Positive Psychology. *AI Practitioner*, 17(3).

Whitney, D., Trosten-Bloom, A., & Vianello, M. G. 2019. Appreciative inquiry: Positive action research. In O. Zuber-Skerritt & L. Wood (Eds.), *Action learning and action research: Genres and approaches*: 163-177. Bingley: Emerald Publishing Limited.

Whitney, D. D. & Trosten-Bloom, A. 2010. *The power of appreciative inquiry: A practical guide to positive change*: Berrett-Koehler Publishers.

Williamson, O. E. 1999. Strategy research: Governance and competence perspectives. *Strategic Management Journal*, 20(12): 1087-1108.

Winnard, J., Adcroft, A., Lee, J., & Skipp, D. 2014. Surviving or flourishing? Integrating business resilience and sustainability. *Journal of Strategy and Management*, 7(3): 303-315.

Yigitcanlar, T., O'connor, K., & Westerman, C. 2008. The making of knowledge cities: Melbourne's knowledge-based urban development experience. *Cities*, 25(2): 63-72.

Yigitcanlar, T. 2010. *Rethinking sustainable development: Urban management, engineering, and design*: IGI Global.

Yigitcanlar, T., Dur, F., & Dizdaroglu, D. 2015. Towards prosperous sustainable cities: A multiscalar urban sustainability assessment approach. *Habitat International*, 45(Part 1): 36-46.

Yigitcanlar, T. & Kamruzzaman, M. 2015. Planning, development and management of sustainable cities: A commentary from the guest editors. *Sustainability* 7(11): 14677-14688.

Yin, R. K. 1994. Discovering the future of the case study method in evaluation research. *Evaluation Practice*, 15(3): 283-290.

Yoo, Y., Boland, R. J., Jr., & Lyytinen, K. 2006. From organization design to organization designing. *Organization Science*, 17(2): 215-229.

Zeng, J., Zhang, W., Matsui, Y., & Zhao, X. 2017. The impact of organizational context on hard and soft quality management and innovation performance. *International Journal of Production Economics*, 185: 240-251.

Zhou, J. & George, J. M. 2001. When job dissatisfaction leads to creativity: Encouraging the expression of voice. *Academy of Management Journal*, 44(4): 682-696.